Policies for Seismic Safety: Elements of a State Governmental Program

INSTITUTE OF GOVERNMENTAL STUDIES

Eugene C. Lee, *Director*

The Institute of Governmental Studies was established in 1919 as the Bureau of Public Administration, and given its present name in 1962. One of the oldest organized research units in the University of California, the Institute conducts extensive and varied research and service programs in such fields as public policy, politics, urban-metropolitan problems, and public administration. The Institute focuses on problems and issues confronting the government and citizens of the San Francisco Bay Area, of California, and of the nation.

The professional staff includes faculty members holding joint Institute and departmental appointments, research specialists, librarians and graduate students. In addition the Institute encourages policy-oriented research and writing efforts by a variety of faculty members and researchers not formally affiliated with the staff. The Institute is also host to visiting scholars from other parts of the United States and many foreign nations.

A prime resource in its endeavors is the Institute Library, with more than 370,000 documents, pamphlets and periodicals relating primarily to government and public affairs. Holdings include a number of major special collections. The Library serves faculty and staff members, students, public officials and other interested citizens.

The Institute also publishes books, monographs, bibliographies, periodicals, research reports, and reprints for a national audience. It issues the bimonthly Institute bulletin, the *Public Affairs Report* dealing with problems in public policy; and the occasional publication, *California Data Brief*, which presents timely data on the state's social and economic development.

In addition, the Institute sponsors lectures, conferences, workshops and seminars that bring together faculty members, public officials, and other citizens. It administers the California Policy Seminar and California Research Colloquium, Universitywide programs designed to promote and facilitate research on public-policy issues confronting the state.

These programs and publications are intended to stimulate thought, research and action by scholars, citizens and public officials on significant governmental policies and social issues.

Policies for Seismic Safety: Elements of a State Governmental Program

STANLEY SCOTT
Institute of Governmental Studies

INSTITUTE OF GOVERNMENTAL STUDIES
UNIVERSITY OF CALIFORNIA, BERKELEY
1979

Library of Congress Cataloging in Publication Data

Scott, Stanley, 1921-
Policies for seismic safety.

Bibliography: p.
1. Disaster relief–Planning–California.
2. Earthquakes–California. 3. Disaster relief–
Planning–United States. 4. Earthquakes–United
States. I. Title.
HV555.U62C37 353.9'794'0075 79-19189
ISBN 0-87772-268-4

Contents

Table

Foreword

Many ad hoc groups have been created to consider earthquake-related matters in California as well as nationally. To set the present report in perspective, it may be useful to note briefly some earlier efforts, most of which were never intended to be continuing entities. One excellent early example was the California State Earthquake Investigation Commission formed in 1906 by Governor George C. Pardee. The commission was headed by Professor A.C. Lawson of the University of California, Berkeley, and was funded by the Carnegie Institution of Washington. It produced a classic, monumental study of the San Francisco earthquake of April 18, 1906, but this commission went out of existence at the study's completion.

Other limited or specific actions were taken in response to subsequent California earthquakes. One of the most notable was the legislative effort shortly after the Long Beach earthquake of 1933, when the Field Act was passed setting the framework for earthquake-resistant standards for public school buildings, and providing for state review and inspection to enforce the law.

The great Alaskan earthquake of 1964 stimulated a renewal of California concern with seismic hazards. In the late 1960s, growing interest within the University community at Berkeley produced an ad hoc group that explored some of the relationships among science, engineering and public policy for earthquake safety. Personnel associated with the Institute of Governmental Studies were instrumental in bringing matters to a focus, with meetings attended by a representative of State Senator Alfred E. Alquist, as well as faculty members and non-University experts concerned with seismic safety.

These efforts led in 1969 to creation of the California Legislature's Joint Committee on Seismic Safety, headed by Senator Alquist. After a four-year study—and some notable legislation on hospital safety and fault zoning passed after the San Fernando earthquake of 1971—the joint committee issued a landmark report on seismic safety needs, and urged creation of a state-level body to continue the work on hazard reduction.* The present California Seismic Safety Commission came into being in 1975, and for the first time California had a continuing state organization with the principal mission of developing public policy and making recommendations with respect to earthquake safety.

Meanwhile at the national level, studies in the middle and late 1960s were urging the development of federal earthquake-hazard mitigation activities, principally through greater support for research related to earthquake safety. These reports stimulated increasing efforts on the part of several federal agencies, but much more attention was still needed. Finally, with the passage of the "Earthquake Hazard Reduction Act of 1977," PL 95-124 (often cited as the Cranston Bill), the federal government has taken strong steps to promote seismic safety in all earthquake-prone areas of the United States.

The 1977 federal legislation is much more broadly based than previous non-legislative efforts, because Congress has thus joined with the Executive Branch in pressing for earthquake-hazard mitigation and improved seismic safety. The federal approach follows the pattern established by California's Joint Legislative Committee on Seismic Safety and the Governor's Earthquake Council (1972-1974). The federal law, in effect, provides for a national counterpart group to California's Commission. The objective of PL 95-124 is to mount a broad frontal attack on earthquake hazards, including:

1. Developing feasible design and construction methods for areas of seismic risk, in order to make new and existing structures earthquake-resistant, giving priority to nuclear power generating plants, dams, hospitals, schools, public utilities, public safety structures, high occupancy buildings, and similar other occupancies.

2. Implementing systems for predicting damaging earthquakes in all areas of moderate or high seismic risk.

3. Developing model safety codes in conjunction with state and local officials, and professional organizations.

4. Improving understanding of earthquakes, and of capability with respect to earthquake safety, disseminating earthquake warnings, organizing emergency services, and planning for reconstruction and redevelopment following earthquakes.

5. Educating the public, including state and local officials, on the significance of earthquakes and related seismic and geologic events.

6. Encouraging research on:

   a. Ways to increase the use of existing scientific and engineering knowledge.

---

*California, Legislature, Joint Committee on Seismic Safety, *Meeting the Earthquake Challenge* (January 1974).

b. The social, economic, legal, and political consequences of earthquake prediction.

c. Ways to improve the availability of earthquake insurance or some functional substitute.

The President assigned responsibility for developing an implementation plan for the federal legislation to the Office of Science and Technology Policy (OSTP). Implementation questions were addressed in the OSTP report *Earthquake Hazards Reduction: Issues for an Implementation Plan* (1978). Several working and advisory groups participated, comprising persons drawn from a broad spectrum of disciplines, including physical sciences, engineering and social sciences, as well as governmental and private-sector policymakers. Consultant organizations and individuals were asked to prepare a variety of special reports, many being based on California's experience. For example, the initial version of this monograph was one such report that tried to interpret the California experience for the benefit of the national effort. Originally prepared a year and one half ago, the paper was subsequently developed and further amplified into its present form. In addition to consultants, more than 70 state and local government, professional, trade, and volunteer organizations contributed to the federal deliberations.

In short, as suggested earlier, earthquake hazard mitigation exemplifies an on-going cyclical, mutually supportive process involving both state and federal levels. Thus the California Seismic Safety Commission and its activities provided examples for use by the federal counterpart. In developing the federal program, the Office of Science and Technology Policy, in turn, commissioned reports that relied heavily on findings based on state experience. Moreover, the new federal legislation presumes that much of the actual policy formulation and enforcement will be done at the state and local level, recognizing that hazard reduction must be a shared responsibility. Important state-level initiatives are already being taken. In addition to the California Commission, Utah legislation established a Seismic Safety Advisory Council in 1977, Nevada has a gubernatorially appointed Ad Hoc Panel on Seismic Hazard Mitigation, and Montana is considering creation of a similar body.

California's seismic safety efforts have been pioneering in several respects, its national and worldwide leadership being acknowledged in many circles. In March 1979, for example, the California Commission, together with the groups from Utah and Nevada and representatives from several other western states, met in Reno, Nevada to discuss mutual concerns and programs. At that meeting a "Western States Seismic Safety Council" was formed to facilitate communication and cooperation among these states.

This timely report distilling the principal results of California's experience for a policy oriented audience is appropriately being published by the Institute of Governmental Studies, with its long-standing interest in earthquake safety policy. The information should be particularly valuable to other states considering seismic safety needs, and the discussion may also be of help in the study of policies related to other natural hazards.

Karl V. Steinbrugge
Working Group on Earthquake Hazard Reduction
Office of Science and Technology Policy
Executive Office of the President
Washington, D.C.

Acknowledgments

The author wishes to acknowledge the advice and help of a large number of people with varied backgrounds who generously read earlier drafts of the entire manuscript or major sections of it. Without their help and substantive contributions this monograph would not have been possible. Beyond those whose names are mentioned below, the author also wishes to give credit to the many other sources drawn upon for ideas about policies for geologic and seismic safety. Accordingly the "Bibliography" attempts to present a reasonably complete listing of the principal written materials used, without claiming to be definitive.

Ayres, Marx
(former member of the California Seismic Safety Commission)
President
Ayres Associates
Los Angeles, California

Barrish, Jack S.
J.S. Barrish, Civil & Structural Engineers
Sacramento, California

Blume, John A.
President
J.A. Blume & Associates
San Francisco, California

Bolt, Bruce A.
Director
Seismographic Stations
University of California, Berkeley

Buck, Richard
Staff
California Seismic Safety Commission
Sacramento, California

Cheesebrough, Fred W.
Chief Structural Engineer
Office of State Architect
Sacramento, California

Cluff, Lloyd
Woodward-Clyde Consultants
San Francisco, California

Cochran, Gilbert F.
Research Professor
Water Resources Center
Desert Research Institute
University of Nevada
Reno, Nevada

Condon, Emmet
(member of the California Seismic Safety Commission)
Deputy Fire Chief
San Francisco Fire Department
San Francisco, California

Cortright, Clifford J.
Consulting Civil Engineer
Sacramento, California

Davis, James F.
State Geologist
Division of Mines & Geology
Sacramento, California

Degenkolb, Henry J.
(former Vice Chairman, California Seismic Safety Commission)
President
H.J. Degenkolb & Associates
San Francisco, California

Degenkolb, Oris
Senior Bridge Engineer
California Department of Transportation
Sacramento, California

Donovan, Aileen
Associate Librarian
EERC Library
University of California, Berkeley

Doody, James J.
Chief
Division of Safety of Dams
Department of Water Resources
Sacramento, California

Dukleth, Gordon
Chief, Division of Design and Construction
Department of Water Resources
Sacramento, California

Ford, Charles
(member of the California Seismic Safety Commission)
San Mateo, California

Gay, Thomas
Chief Deputy State Geologist
Resources Agency
Division of Mines and Geology
Sacramento, California

Giersch, Louise
(Vice Chairman, California Seismic Safety Commission), Regional
Director, Air & Hazardous Materials Division, Environmental Protection Agency

Gilbertson, C.L.
Administrator
Disaster & Emergency Services Division
Helena, Montana

Hart, Earl W.
Senior Geologist
Division of Mines and Geology
San Francisco, California

Jahns, Richard H.
(member of the California Seismic Safety Commission)
Dean, School of Earth Sciences
Stanford University
Stanford, California

Lagorio, Henry J.
(former member of the California Seismic Safety Commission)
Director of Research
Department of Architecture
University of Hawaii at Manoa
Honolulu, Hawaii

Mader, George
(member of the California Seismic Safety Commission)
Planning Consultant
William Spangle & Associates
Portola Valley, California

Mann, Arthur
(former member of the California Seismic Safety Commission)
Solvang, California

McClure, Frank E.
Plant Engineer
Lawrence Berkeley Laboratory
University of California, Berkeley

McKenzie, Richard
Seismographic Stations
University of California, Berkeley

McLeod, John
Staff
California Seismic Safety Commission
Sacramento, California

Meehan, John F.
Principal Structural Engineer
Office of State Architect
Structural Safety Section
Sacramento, California

Moore, Michal
(member of the California Seismic Safety Commission)
Monterey County Board of Supervisors
Monterey, California

Olson, Robert
Executive Director
Seismic Safety Commission
Sacramento, California

Perry, Will
(member of the California Seismic Safety Commission)
Director, Contra Costa County Office of Emergency Services
Martinez, California

Pulley, H. Roger
Senior Planner
Office of Emergency Services
Sacramento, California

Rigney, Robert
(Chairman, California Seismic Safety Commission)
Administrator
Environmental Improvement Agency
San Bernardino County, California

Russell, Katherine
(member of the California Seismic Safety Commission)
Board and Executive Committee Member
Los Angeles Chapter of the American National Red Cross
Los Angeles, California

Seed, H. Bolton
(member of the California Seismic Safety Commission)
Professor
Department of Civil Engineering
University of California, Berkeley

Shah, Haresh C.
Director, The John A. Blume Earthquake Engineering Center
Stanford University
Stanford, California

Slosson, James E.
(former member of the California Seismic Safety Commission)
Chief Engineering Geologist
Slosson & Associates
Van Nuys, California

Steinbrugge, Karl V.
(member and former Chairman, California Seismic Safety Commission)
Manager
Earthquake Department, Pacific Region Insurance Services Office
San Francisco, California

Stromberg, Peter
Staff, California Seismic Safety Commission
Sacramento, California

Thiel, Charles C., Jr.
Problem Focused Research Applications
National Science Foundation
Washington, D.C.

Wallace, Robert E.
Chief Scientist
Office of Earthquake Studies
United States Geological Survey
Menlo Park, California

Ward, Delbert B.
Executive Director
Seismic Safety Advisory Council
Salt Lake City, Utah

Wyner, Alan
Associate Professor
Department of Political Science
University of California
Santa Barbara, California

Introduction: Seismic Hazards in California and the Nation

Damaging earthquakes are fairly frequent in many parts of the western United States. The historic record includes several large ones in the 1800s, and the devastating quake and fire that destroyed much of San Francisco in 1906 firmly established California as "earthquake country" in the eyes of the nation and the world.

Since then, many smaller but damaging earthquakes have occurred in California and other western states, and in 1964 Alaska was hit by a huge quake of even larger magnitude than the 1906 event. These seismic events have produced reasonably realistic public awareness in the western United States, especially California, that future earthquakes are inevitable, and that precautions and advance preparations are very important if the high potential for casualties and damage is to be reduced.

A Great California Earthquake Imminent?

Many experts now believe that a great California earthquake is imminent. This conclusion is based on the historic record of large earthquakes, on recent research into prehistoric earthquakes, and on knowledge of plate tectonics, the movement of large slabs of the earth's lithosphere past one another.

The available historic record shows four great earthquakes in Northern California in the 1800s and early 1900s affecting areas now heavily urbanized. These occurred in 1836 (Hayward Fault), 1838 (San Andreas Fault), 1868 (Hayward Fault), and 1906 (San Andreas Fault). Southern California experienced one great historic earthquake centered near Ft. Tejon, between Bakersfield and Los Angeles, in 1867 (San Andreas Fault).* Moreover recent geologic

*There was also a very great earthquake in Owens Valley in 1872, but this site is relatively distant from intensively urbanized areas.

studies have extended the record back in time, identifying a series of great prehistoric earthquakes in Southern California whose recurrence intervals vary considerably but appear to average about 200 years.

Finally, boundaries between major continental plates are known to pass through California, and these plates are moving with respect to one another. Plate tectonics is in fact the most convincing and currently accepted explanation for a majority of earthquakes wherever they occur. Plate movements in California are demonstrated by such evidence as the northwesterly drift—approximately three inches per year—of the Farallon Islands, about 30 miles west of San Francisco, with respect to Mt. Diablo, 30 miles east of San Francisco. The resulting strain building up in the intervening formations will have to be relieved by slippages that will almost certainly occur along one or more of the major active faults traversing the San Francisco Bay Area. Given the long 73-year interval since the most recent great earthquake in Northern California, the amount of slippage that must occur will be sufficient to produce one or more great earthquakes.

Relying on such evidence, University of California seismologist Bruce A. Bolt recently put forth a "working hypothesis" that a great earthquake is likely to strike somewhere in California within the next ten years, and probably will affect some major urban areas. Bolt estimates the likelihood of this happening within ten years as higher than 50 percent. Moreover he points out that the probability of such an earthquake occurring within a specified period increases progressively, as more time elapses since the last great earthquake.

East of the Rockies

The situation is different in many other parts of the nation, particularly east of the Rocky Mountains. Earthquakes are far less frequent there than in the western United States, including Alaska and Hawaii. The first three maps illustrate these contrasts in different ways. Map 1 shows the number of earthquakes with Modified Mercalli intensities of IV or greater in a recent 45-year period.* Map 2 gives earthquake intensities of V and above observed from the beginning of effective records through 1970.

Both maps show (1) the far greater seismicity of the western area, and (2) the lower but still appreciable seismicity east of the Rockies. In a slightly different way, Map 3 also demonstrates the seismicity found in the Midwest and

---

*Appendices A and B give brief explanations of the two principal systems of earthquake measurement, the logarithmic Richter scale which expresses earthquake *magnitudes*, and the Modified Mercalli scale used to record *intensities* of observed earthquake damage and other effects.

Map 1 Number of Earthquakes with Intensities of IV or Greater Experienced from 1928 through 1973

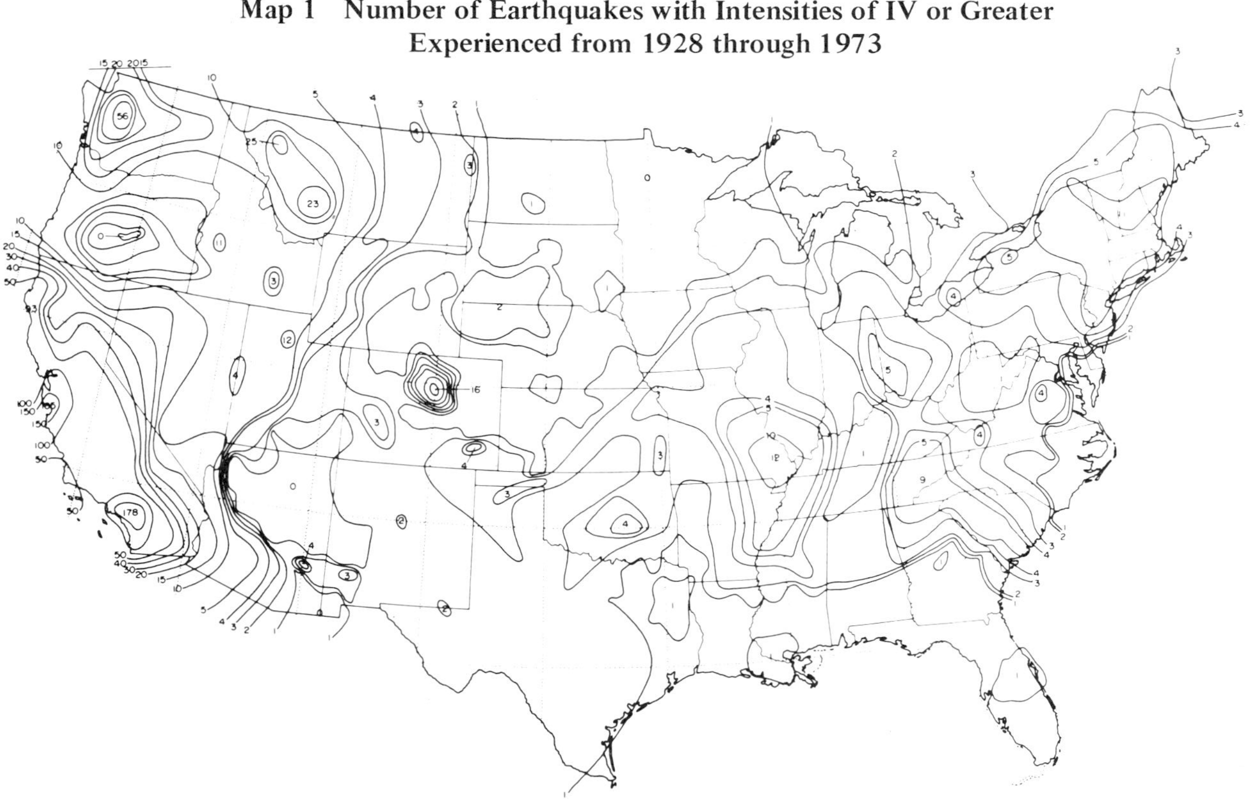

Source: National Oceanic and Atmospheric Administration (from R.J. Brazee, *An Analysis of Earthquake Intensities . . .* August 1976).

Map 2 Earthquakes with Intensities of V and above in the United States (through 1970)

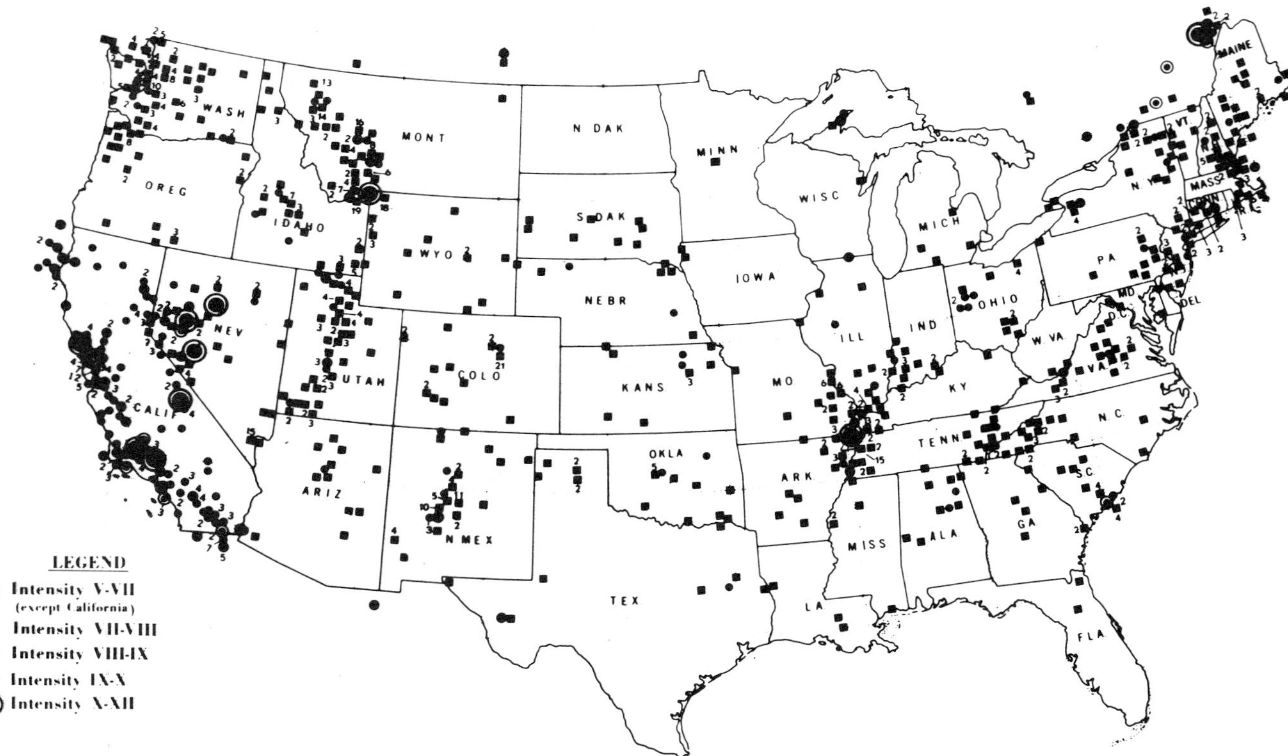

Notes: Because of their large number, California earthquake intensity observations of V up through VII are not shown. The small Arabic numbers indicate the number of earthquake intensity observations made at a particular site.

Source: National Oceanic and Atmospheric Administration (from Jerry L. Coffman and Carl A. von Hake, *Earthquake History of the United States*, 1973).

Map 3 Maximum Earthquake Intensities Experienced from 1928 through 1973

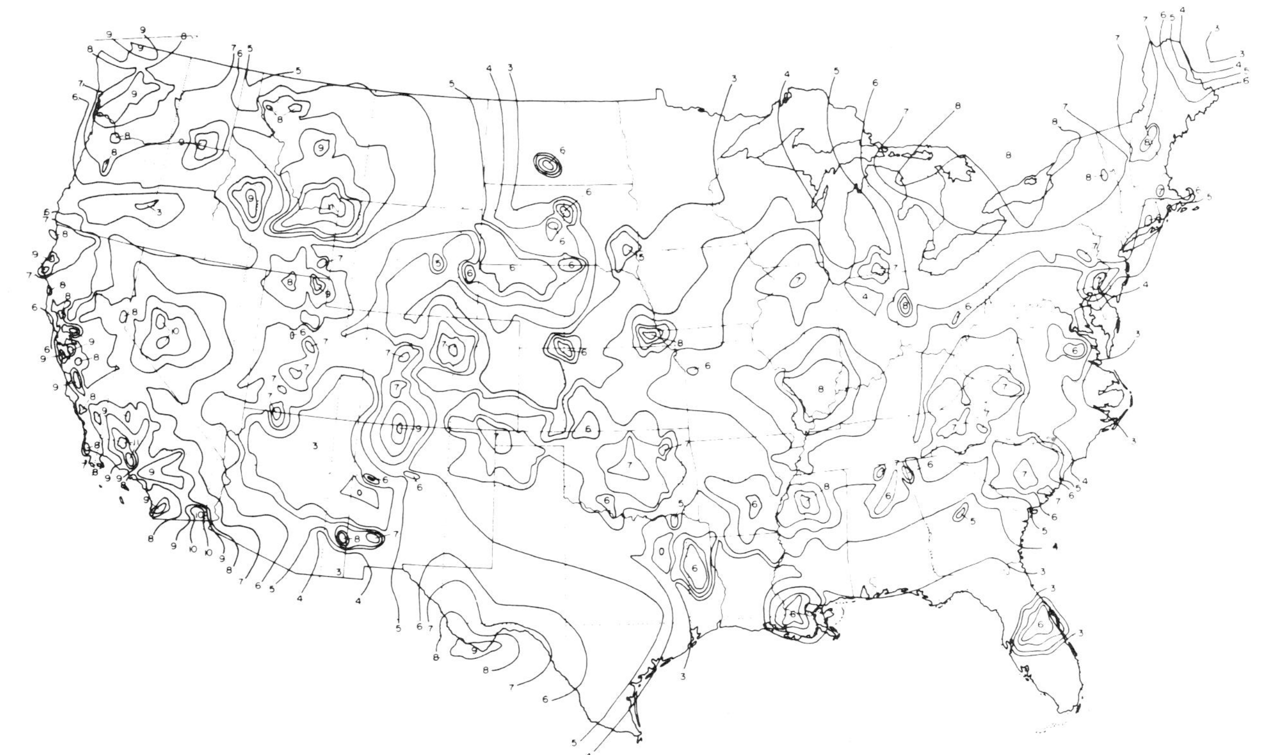

Note: For convenience, the maximum intensity contours mapped here are designated by Arabic numerals rather than the Roman numerals conventionally used to express earthquake intensities.

Source: National Oceanic and Atmospheric Administration (from R.J. Brazee, *An Analysis of Earthquake Intensities* . . . August 1976).

East. Together, Maps 2 and 3 show that intensities from V through VIII have been observed in many parts of the United States east of the Rockies.*

In short, the earthquake hazard east of the Rockies is far less than in the western United States, but by no means negligible. Major earthquakes of Richter magnitudes 7.3 to 7.8 occurred near New Madrid, Missouri in 1811 and 1812, and were felt over large regions of the country (see Map 4). Charleston, South Carolina suffered heavy damage and many casualties in an 1886 earthquake that was felt as far away as Boston, Chicago and St. Louis.

Comparable earthquakes or perhaps even larger ones could occur in the eastern and midwestern United States, although present knowledge of the region's seismicity is too limited to provide reliable indications as to where or when. Smaller but damaging earthquakes occur a good deal more often than the big ones just noted, as we have already seen from Map 1 and Map 2.

Seismic Probability Comparisons

Map 5 takes a different approach in presenting current estimates of regional seismic probabilities. The contours show the effective peak acceleration of the ground—expressed as a fraction of the pull of gravity at the earth's surface—that can be expected, with odds of only one in ten that these peaks will be exceeded in any 50 year period.** Once more, the greater seismicity of the western United States comes through clearly, as does evidence of substantial seismicity in many eastern and midwestern areas. But interpretations of Map 5 must also be carefully qualified, because the underlying causes of seismic activity east of the Rockies are poorly understood, leaving much uncertainty about the recurrence of damaging earthquakes and their probable future locations. In fact, because its non-western portions were constructed with little knowledge of the faults that have produced past earthquakes, the map may require thoroughgoing revision after only one or two major new earthquakes east of the Rockies.

---

*Earthquake damage of consequence begins at intensity VII and gets progressively worse as one goes up the scale. Thus an intensity of VII means considerable damage to badly designed or poorly built structures, while others of moderate-to-good design will experience only slight or negligible damage. An intensity of VIII means great damage in poorly built structures, and considerable damage including partial collapse in ordinary substantial buildings, but only slight damage in structures especially designed to resist earthquakes. See Appendix B.

**Average or effective ground peak acceleration estimates are important because they have a bearing on forces affecting structures during earthquakes, and are employed by engineers in making decisions on building design.

Map 4 Isoseismal Lines of Damage Intensity in the December 16, 1811 Earthquake Near New Madrid, Missouri

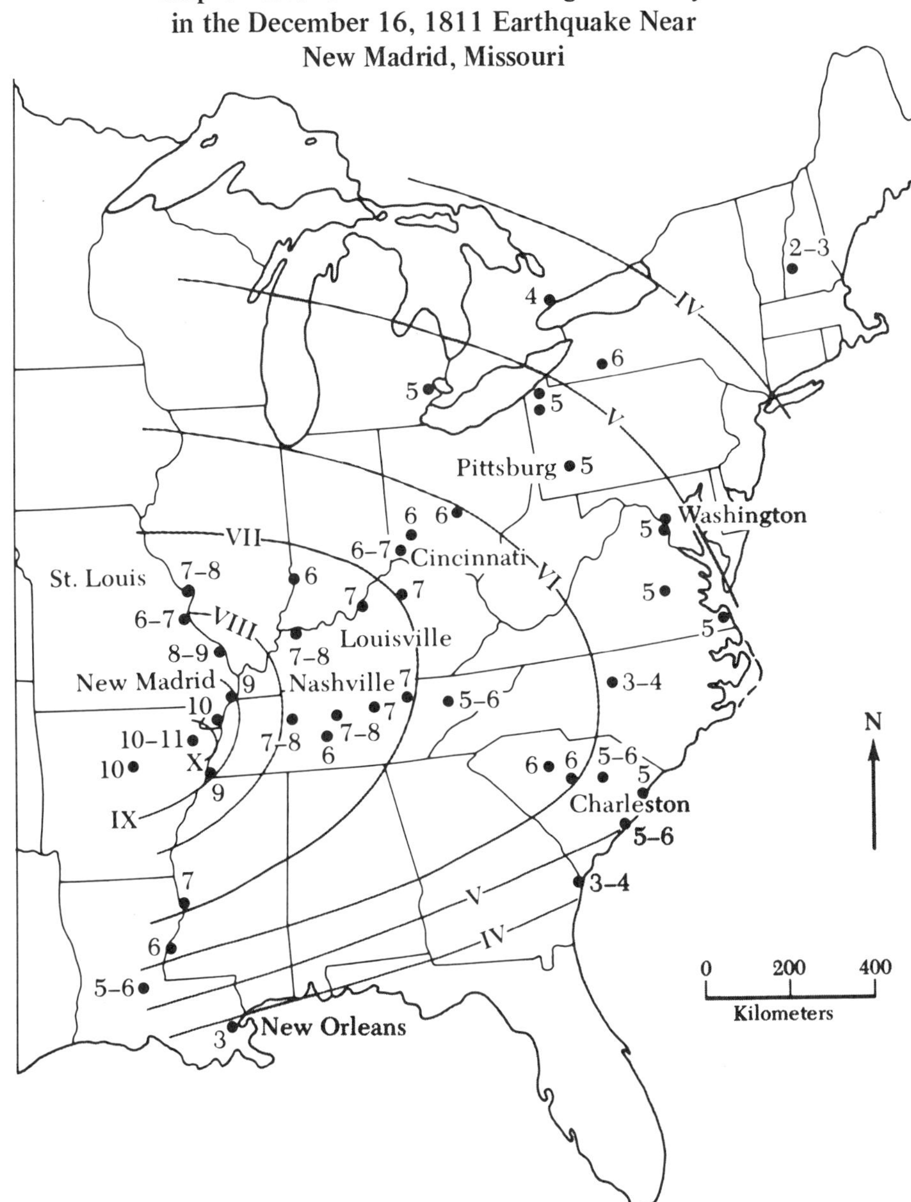

Note: Intensity values at specified points are given in Arabic numerals, while the isoseismals (lines of equal intensity) are labeled by Roman numerals.

Source: Bruce A. Bolt, *Earthquakes: A Primer* (1978), p. 101 (courtesy of O. Nuttli and *Bull. Seism. Soc. Am.*).

Map 5 A Seismic Risk Map for the United States

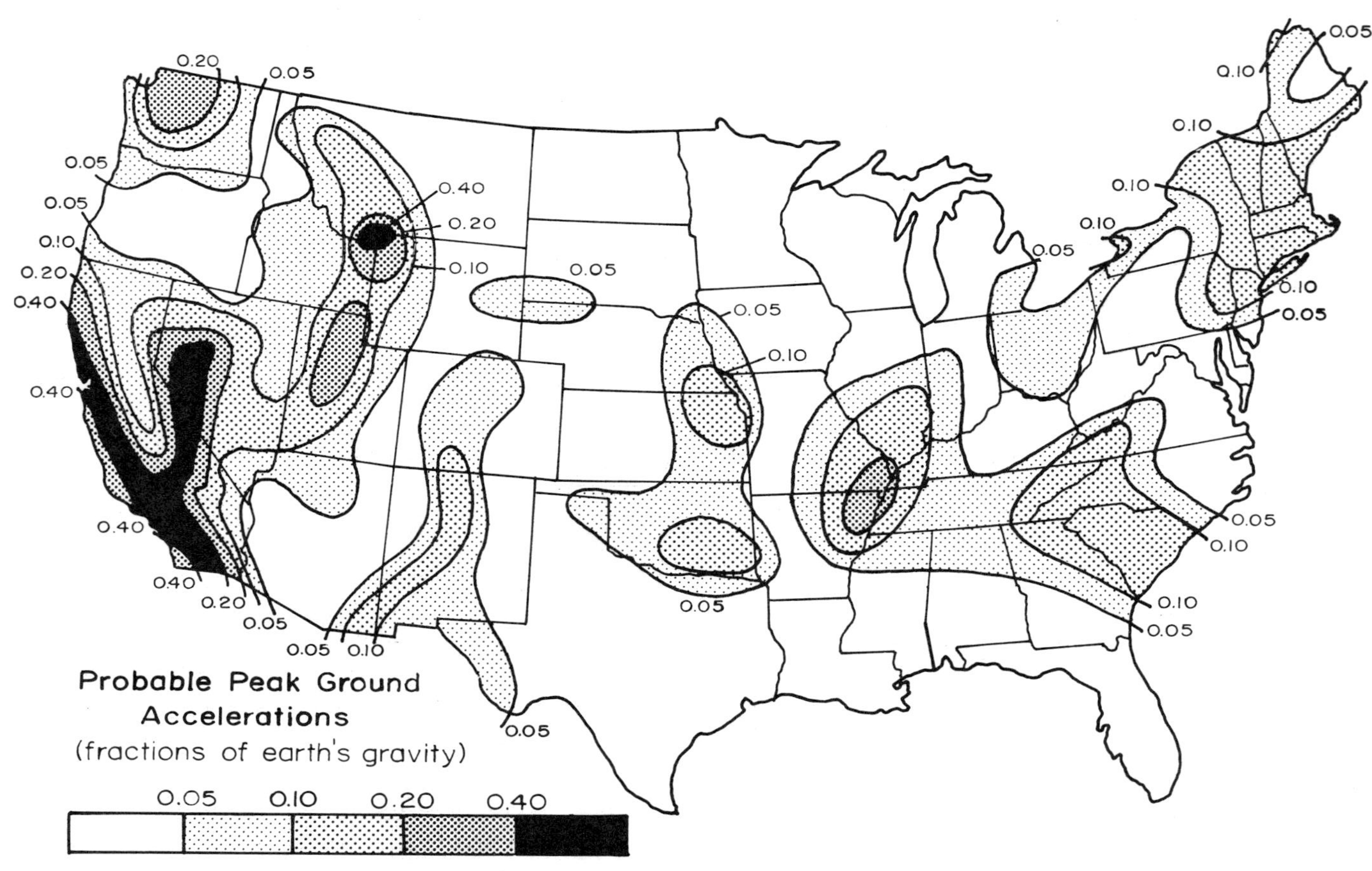

Geologist Clarence Allen, California Institute of Technology, points out that even in California and Nevada, where the historical record is about a century and a half long, "we must be exceedingly cautious in extrapolating from this very short history." He continues with respect to the eastern United States:

> The problem gets even more difficult as we get farther and farther away from active plate boundaries and into areas of low long-term seismicity. What conclusions, for example, can be drawn from the single great earthquake at Charleston, South Carolina, in 1886? Is Charleston really any more dangerous in terms of another similar earthquake tomorrow than. . .Washington, D.C. or New York City? The single historical event tells us essentially nothing in itself except that earthquakes of the same magnitude must therefore be considered credible events, however unlikely, throughout the same entire tectonic province, at least until we under-

---

Note to Map 5: Map 5 attempts to show graphically the likelihood, based on the historic record of past earthquakes, that ground shaking of different severities will occur in the continental United States. The contours indicate the effective peak (maximum) acceleration levels, expressed in decimal fractions of earth's gravity, that have only a 10 percent chance of being exceeded during a 50-year period.

Because large earthquakes are much more frequent in the West, the map shows that region's hazard as clearly much greater than in other regions. On the other hand, the map's non-western portions are not definitive, being based on scarce available information about the faults that have produced past earthquakes. Because the underlying causes of earthquakes east of the Rockies are poorly understood, judgments as to probable frequencies in that region are highly speculative and subject to change as more is learned about earthquake mechanisms east of the Rockies. Readers should be aware of these limitations, otherwise Map 5 can be misleading, possibly disposing some eastern and midwestern users to discount earthquake hazards altogether. This would be a major mistake, as damaging earthquakes have occurred east of the Rockies and will again.

Further, as noted in the text, when substantial earthquakes do strike east of the Rockies, they can be much more destructive than earthquakes of similar magnitudes in the West. Thus it has been estimated that the total area in the non-western United States that has experienced destructive ground shaking at some time during the past 250 years is probably greater than in the West.

Sources: Based on the work by the Applied Technology Council, associated with the Structural Engineers Association of California. Much of the statistical work on risk maps was done by seismologists in the U.S. Geological Survey, especially S.T. Algermissen and David Perkins. (From Bruce A. Bolt, *Earthquakes: A Primer* (1978), pp. 174-176.)

> stand from geological and geophysical studies why the Charleston earthquake occurred where it did and how other areas are truly different.*[1]

Geophysicist B.F. Howell, Jr., notes that seismic activity in the East tends to come "in spurts." Writing in 1973, he commented:

> . . .the last 29 years have been a time of less activity than previously in marked contrast to the three earlier roughly hundred-year-long bursts of activity starting in New England in 1727, New Madrid, Missouri in 1811, and Charleston, South Carolina in 1886. Intensity 8 or larger earthquakes have occurred in the east on the average four times in every 50 years for the past 200 years. The next one is overdue. If past history is a guide, there is a considerable likelihood that it will occur at a place which has not previously experienced such an earthquake.[2]

Readers should be aware of these limitations on our present knowledge, and it must be recognized that our record of historic seismicity in the United States is too short to serve as an accurate measure of seismic risk in all regions. As Map 5 indicates, damaging earthquakes can be expected to strike occasionally in many parts of the nation, including the East and Midwest, and no part can be presumed wholly free from seismic hazard in the long run.[3]

Uncertainties Summarized

Engineering geologist Lloyd Cluff sums up some of the uncertainties that make it difficult to assess earthquake hazard in the East, focusing on information about fault activity:

---

*The very lack of reliable information leads various experts to different conclusions when speculating on the potential for seismic activity in regions east of the Rockies. For example, University of California seismologist Bruce A. Bolt comments that he "disagrees substantially" with the implications of the last two sentences in Allen's comments quoted here. On this point, Stanford University geologist Richard H. Jahns notes that, while Allen's conclusion may be open to question, partly because his emphasis on "*entire* tectonic province" may be "too much," he is protected by use of the word "credible." In other words great earthquakes in those regions "cannot be considered *in*credible." Jahns concludes: "It's mainly a matter of what one wishes to emphasize."

> Active faults are more easily recognizable in the west than in the east because of several factors: (1) the most important . . .is that the degree of fault activity is much higher along many faults in the west, thereby resulting in more abundant evidence both in the geologic record and the historical record. (2) In large areas of the eastern United States, there is an absence of geologically recent deposits that would record evidence of recent fault activity. (3) The heavy vegetative cover in the eastern United States makes it difficult to identify faults that would otherwise be obvious. (4) Techniques developed over the past few years in the west for evaluating fault activity have not been applied in the east. . .[4]

A Western Advantage: Attenuation of Earthquake Forces

Comparisons of regional seismic hazards must take account of another important difference: Earthquakes of equal magnitude cause destruction over wider areas east of the Rockies than in the western United States. This is due to the greater attenuation or reduction of earthquake waves as they pass through typical western geologic formations. Attenuation in the East is several orders of magnitude lower, permitting earthquake forces to be transmitted farther, and to shake much larger areas, perhaps severely, thus doing more widespread damage.[5] The ground materials often found east of the Rockies may also be more likely to magnify shaking and thus increase damage. Because of these effects, for example, the three New Madrid earthquakes mentioned earlier—which were substantially smaller than the 8.3 magnitude San Francisco quake of 1906—produced damage rated "strong" out to points about 500 miles from New Madrid. In contrast, strong damage extended only about 60 miles from the center of the great San Francisco quake.

To sum up, the regions east of the Rockies have an advantage over the West in experiencing far fewer earthquakes. But the eastern U.S. also has the disadvantages of wider areas probably being damaged by those large earthquakes that do occur, and greater uncertainty as to when and where earthquakes are likely to strike.

California's Greater Recognition and Response

The eastern U.S. has another disadvantage, because seismic hazard has received only limited public recognition there. Consequently in most areas of the East and Midwest little if any attention has been given to the need for safety measures and precautions that would mitigate future earthquake damage. As a result the region has a high proportion of structures that were not built to be earthquake resistant.

By contrast, California's recognition of earthquake hazard has produced several policy responses, including greater concern with earthquake-resistant design, state seismic safety measures relating to public schools and hospitals, restrictions on building across faults, seismic planning by local governments, and improved dam safety procedures. Perhaps the most important action was creation of a state Seismic Safety Commission in 1975. While a few other western states have shown some response to earthquake hazard, so far none has approached California in seismic safety efforts. Much of the discussion in this monograph is, in fact, based on the California experience, which is described and interpreted for possible adaptation by other states in formulating their own seismic safety or natural hazards programs.

These comments on California's experience call for a few words of caution. Their state's demonstrated leadership in seismic safety gives Californians little cause for self-congratulation. After all, this state has much greater problems, hence a greater need for leadership. Moreover those who are most knowledgeable about the requirements of seismic safety agree that California has a long way to go before it will have put enough thought, effort and resources into bringing earthquake preparedness up to an acceptable level.

Policies for Earthquake Safety: Defining Risks and Reducing Hazards

Given the pronounced regional differences in seismicity and geologic formations, it would not be easy and may not be appropriate to formulate uniform policies to mitigate seismic risk across the whole country. At a minimum, however, it seems prudent for relatively inexpensive safety measures to be taken virtually everywhere. Also, regardless of their geographic locations, structures with long useful lives or housing substantial numbers of people ought to be made resistant to levels of seismic shaking that may reasonably be anticipated.

A recent report on hazard assessment concludes:

> . . .probability analyses can serve a very valuable function if they are related to economic risk [depending]. . . on the intended life of a structure and the consequences of its failure. If the consequences of failure are severe, as with a dam in an urban area, one should design for the worst event regardless of its probability of recurrence. Where failure has only economic consequences and an owner wishes to accept some risk in the interest of reduced construction cost, it is appropriate to accept some less conservative design.[6]

Stronger policies will be appropriate in regions of known higher hazards, and for facilities whose damage or destruction would inflict great harm. A basic rule for a seismic region like the one encompassing California would have all structures intended for human use strong enough to withstand large or great earthquakes–approximately 7.0 to 8.5 or higher on the Richter magnitude scale–minimizing the possibilities of buildings collapsing and killing or injuring substantial numbers of people. Policymakers, designers and builders should also consider the fact that smaller earthquakes of 6.0 to 7.0 magnitude—which are much more frequent than the large events—can cause about as much damage locally as the really big ones.*

Reducing Present Hazards

In any event, the high levels of hazard found in many areas and in many types of structures clearly justify careful attention to seismic safety needs. But also important is a realistic understanding that reducing risks in populated seismic regions like California will be difficult and costly. Existing hazards often present very tough problems. For example, it will be expensive to remove the dangers posed by the presence in all major urban areas of buildings lacking earthquake resistance. Moreover, many of the most hazardous buildings are occupied by lower-income populations, minorities and elderly. Some businesses are also housed in substandard and hazardous buildings and may depend on low rents for economic survival.

Existing dangers are not limited to old buildings, as the design and construction of some modern structures may be inadequate to protect occupants and resist earthquakes. This is also likely to be true of several kinds of critical facilities whose failure could affect major populations and large areas. In short, dealing with existing hazards will require careful study, thoughtful ordering of priorities and costs, and diligent implementation.

New Buildings and "Acceptable Risk"

In the case of buildings not yet constructed, increased earthquake safety—or reduced risk— can be attained with today's knowledge much more readily than when reinforcing existing structures. Realistic use of the best current "state of the art" methods to insure adequate earthquake resistance in the design and construction of new buildings is much less costly than reducing excessive hazards in standing structures.

---

*The worldwide average annual frequency of earthquakes, scaled according to magnitude, emphasizes the much greater frequency of 6's or 7's, in contrast to the very large magnitude 8's:

Richter Magnitude	Number
8	2
7	20
6	100

Source: Bruce A. Bolt, *Earthquakes: A Primer* (1978), p. 195.

In any event, achieving safety in both existing buildings and new ones should be guided by a consensus on the degree of risk to life and property that society is willing to accept. Such risk levels must be defined partly in relation to estimates of the deaths, casualties and damage that earthquakes are likely to cause. Realistic risk evaluations must also weigh the costs of alleviating hazards or avoiding new "unacceptable" hazards in future construction.

Reliable data for calculating cost-benefit ratios are unfortunately often unavailable, and many significant costs and benefits cannot be expressed in quantifiable terms. On the other hand, practical guidelines can be developed, such as the categories of "acceptable risks" outlined in Table 1, which was formulated for use in California. The principal criterion for grouping various kinds of structures into the four-level classification is the numbers of people likely to be affected by the impact of a major earthquake on each type of structure. When larger numbers of people are likely to be affected, the level of risk considered acceptable must be reduced.

Table 1 also gives rough estimates of costs, based on the assumption that society can afford and may be willing to bear the expenses required for improved design and construction to achieve the degree of safety (or risk level) indicated as "acceptable" for each kind of structure.

Strengthening Structures and Other Measures

Table 1 does not, however, estimate what it would cost to reduce earthquake hazards in existing buildings. Determining the costs of strengthening hazardous structures will ultimately depend on individual inspections of large numbers of buildings. Special codes for hazardous buildings will also need to be devised, allowing owners to reinforce them sufficiently to prevent death or injury without necessarily making the additional costly improvements needed for full compliance with current codes.

Beyond safety standards for structures and facilities, seismic safety can also be sought through urban planning and other measures governing the conservation, use and development of land. Particular consideration should be given to the safety implications of major developments that may generate urban growth, in order to discourage unsafe construction in areas likely to be geologically hazardous.

Advance planning can guide rehabilitation and redevelopment after earthquake disasters, thus avoiding the creation or perpetuation of hazards when damaged structures are repaired or replaced. Emergency plans for immediate action in and after earthquakes are essential to rescue and care for survivors, control fires, and limit other disasters that may be triggered by earthquakes. In addition, longer-term financial aid can also help with post-earthquake recovery.

Table 1
Classification of Acceptable Risks
(This table was developed for use in California)

Level of Acceptable Risk	Kinds of Structures	Extra Project Cost Probably Required to Reduce Risk to an Acceptable Level
1. Extremely low[1]	Structures whose continued functioning is critical, or whose failure might be catastrophic: nuclear reactors, large dams, power intertie systems, plants manufacturing or storing explosives or toxic materials	No set percentage (whatever is required for maximum attainable safety)
2. Slightly higher than under level 1[1]	Structures whose use is critically needed after a disaster: important utility centers; hospitals; fire, police, and emergency communication facilities; fire stations; and critical transportation elements such as bridges and overpasses; also smaller dams	5 to 25 percent of project cost[2]
3. Lowest possible risk to occupants of the structure[3]	Structures of high occupancy, or whose use after a disaster would be particularly convenient: schools, churches, theaters, large hotels, and other high-rise buildings housing large numbers of people, other places normally attracting large concentrations of people, civic buildings such as fire stations, secondary utility structures, extremely large commercial enterprises, most roads, alternative or noncritical bridges and overpasses	5 to 15 percent of project cost[4]
4. An "ordinary" level of risk to occupants of the structure[3,5]	The vast majority of structures: most commercial and industrial buildings, small hotels and apartment buildings, and single family residences	1 to 2 percent of project cost, in most cases (2 to 10 percent of project cost in a minority of cases)[4]

The range of policies that can be suggested in the interest of earthquake safety emphasizes the complexity of the problem. Putting all the pieces together will require a new level of strategic "overview" planning. Moreover, comprehensive seismic safety programs cannot be planned and implemented successfully without involving the three principal levels of government—federal, state and local.

Notes to Table 1.

1. Failure of a single structure may affect substantial populations.

2. These additional percentages are based on the assumption that the base cost is the total cost of the building or other facility when ready for occupancy. In addition, it is assumed that the structure would have been designed and built in accordance with current California practice. Moreover, the estimated additional cost presumes that structures in this acceptable-risk category are to be sufficiently safe to remain functional following an earthquake.

3. Failure of a single structure would primarily affect only the occupants.

4. These additional percentages are based on the assumption that the base cost is the total cost of the building or facility when ready for occupancy. In addition, it is assumed that the structures would have been designed and built in accordance with current California practice. Moreover, the estimated additional cost presumes that structures in this acceptable-risk category are to be sufficiently safe to give reasonable assurance of preventing injury or loss of life during an earthquake, but otherwise not necessarily to remain functional.

5. "Ordinary risk": Resist minor earthquakes without damage; resist moderate earthquakes without structural damage, but with some non-structural damage; resist major earthquakes of the intensity or severity of the strongest experienced in California, without collapse, but with some structural as well as non-structural damage. In most structures, it is expected that structural damage, even in a major earthquake, could be limited to repairable damage. (Structural Engineers Association of California.)

Source: California, Legislature, Joint Committee on Seismic Safety, *Meeting the Earthquake Challenge* (1974) p. 9.

Governmental Responsibilities: Policy and Implementation

Each level of government—federal, state, and local—has crucial roles to play in the quest for public safety in seismic regions. While each level needs the other's contribution in an intergovernmental partnership, the state governments are clearly in a pivotal position. Accordingly, after this introductory overview, the remainder of the monograph will give principal attention to major safety issues in seismic regions and recommended measures that state government could adopt to advance the cause of earthquake hazard reduction.

The Federal Role

The federal government's ultimate responsibility for the general welfare of its citizens gives it a basic duty to guard against natural disasters and assist in recovery from them. To this end, the national government should play a crucial role in developing effective intergovernmental and private-sector programs to reduce natural hazards, including earthquake hazards.

The Earthquake Hazard Reduction Act of 1977 represents a major national policy commitment to the quest for improved seismic safety (P.L. 95-124, Oct. 7, 1977; 91 Stat. 1098). The new act contemplates a wide array of federal initiatives. For example it acknowledges that a very important federal function is to provide guidance to state and local governments in disaster policy efforts, and to provide fiscal assistance where necessary to the adequate development and effective implementation of disaster programs. Further, the federal government should both stimulate and monitor state-local disaster preparedness programs, and consider the application of federal safety standards as a back-up in the case of states that consistently fall short.

The federal government also has a key responsibility for conducting or funding basic research on earthquake and other natural hazards. Perhaps even more important are federally funded efforts to determine the best and most effective ways of preparing for earthquakes and other disasters, and of implementing safety policies.

The federal government should also formulate plans for national action in case of major disasters, including plans for large-scale responses to major damage or loss of life, rescue and evacuation, and provision of medical care, food, and police protection, as well as other emergency help.

In general, the distribution of responsibilities among governmental levels could be similar to the formulas used for other major projects where federal initiatives have recently been taken, e.g., environmental protection under the federal Environmental Protection Agency, or coastal planning and regulation under the Office of Coastal Zone Management.

State Governments

The state government in an area of known seismic activity has major responsibilities and should play several pivotal roles in the quest for seismic safety. An important but difficult state task is to evaluate hazards and to determine levels of protection and life safety standards that are judged adequate to safeguard its citizens against earthquakes and other natural disasters. Measures can then be taken to help insure reasonable performance in achieving these safety levels. Thus the state should set minimum safety standards for all kinds of structures, and take ultimate direct responsibility for important types of facilities whose failure in an earthquake would affect large numbers of people. Another state responsibility is to identify and evaluate major fault zones and areas of earthquake hazard, and to prepare guidelines for land-use regulation and construction in the zones.

The state should also review the performance of state and local governmental agencies, adopting new seismic safety policies where necessary to insure that performance measures up at least to acceptable minimum levels. It should also take leadership in planning for future disasters and for disaster recovery, and should represent the state's interest at the national level in pressing for federal disaster preparedness and recovery measures, and appropriate joint federal-state programs.

While some state programs can be directly administered by state agencies, many will be joint efforts with local governments. The state can set standards and guidelines under which local governments carry out earthquake safety policies at the county and municipal levels. When relying on local government for implementation, however, experience with seismic safety and other matters argues strongly that state monitoring and back-up measures are essential to insure the consistency and reliability of local performance.

California's experience in providing for the seismic safety of public schools, hospitals, dams and freeways, in regulating construction in fault zones, and in mandating city and county seismic safety plans is reviewed in subsequent sections of this monograph. Other states may find useful guidance in this experience.

To step up its seismic safety efforts and provide a state-level body responsible for reviewing earthquake policy and monitoring its effectiveness, California created the state Seismic Safety Commission in 1975. This was the culmination of a six-year effort begun in 1969 under a joint committee of the state legislature, and accelerated after the damaging San Fernando earthquake of 1971. In addition to its other activities, in 1978-79 the commission prepared and adopted a draft of long-range seismic safety policies to guide state and local implementation efforts. Other states considering seismic safety needs and policies may find the policies of the California commission of considerable value. (See "Conclusion," pp. 58-61, for further comment on the activities of the Seismic Safety Commission, including its overview report on all the state's seismic safety programs, and its work on long range policies for California.)

Local Governments

Local governments normally carry out many police power regulations whose effective enforcement is essential to public safety, including seismic safety. Thus local government is recognized as responsible for enforcing most building and construction code requirements and land-use regulations, as well as providing many water, sanitation, and other forms of utility and "lifeline" services. In exercising these powers, local agencies should take important responsibilities for immediate responses after disasters, as well as for sound longer-term planning for future disasters, and for appropriate preventive measures.

If their authority is properly used, local governments can implement most of the seismic safety regulations required for the health and welfare of their citizens, but will also need appropriate state guidance, encouragement and assistance. In addition to providing such assistance, the state government should monitor and evaluate local efforts, and consider new legislation or other measures necessary to help local government carry out its crucial responsibilities for minimizing casualties and losses from future earthquakes and other natural disasters.

Finally, the state government must also clearly recognize that some disasters are beyond the capabilities of local government to prepare for or cope with. Where this is the case—and it will be with most earthquake disasters—the state should play a strong role in preparing for future natural disasters considered likely to strike.

Regulating the Design and Construction of Buildings

Probably a state's most important single responsibility for natural hazards is to take practicable measures to assure reasonable performance of structures when they are subjected to natural catastrophes. With respect to earthquakes, as noted earlier, a state's principal means of providing for the seismic resistance of new structures is to see that building codes and safety standards are adequate, and that they are enforced with reasonable consistency.

Formulating Appropriate Codes and Standards

A previous chapter discussed policies for seismic safety, including the need to evaluate hazard levels and the probability of future earthquakes in formulating criteria and standards for acceptable risks. States should play a principal role in these evaluations, acting on the best information and advice obtainable, assisted by appropriate agencies of the United States government, and by qualified private-sector experts.

Codes and standards, especially those for lateral forces (sideways shaking) need to be revised and refined as more is learned about earthquake mechanisms, and about the performance of structures subjected to various kinds of movement. Thus state governments should improve regulations as needed, in cooperation with professional societies and institutions of higher education. A state seismic safety or natural hazards agency should review proposed code revisions and recommend the adoption of new codes or new state seismic safety policies as needed.

Enforcement and Implementation

Probably a crucial weak point in seismic safety policy is the enforcement of seismic design regulations. The comparatively low salaries typically paid inspection personnel by local government are an important obstacle to good performance. Accordingly state governments should do what they can to assure the availability of specialized personnel competent in seismic safety and familiar with sophisticated forms of earthquake-resistant construction.

The adequacy of local design review and enforcement should be monitored, and the performance of local building departments in reviewing and approving construction plans upgraded. For this, the states may need to provide for supplementary qualified personnel, where local staff are not adequate. Expert seismic review panels could be established to assist with the monitoring and upgrading, as well as to participate in reviews of special, unusual or critical facilities. A state seismic safety or natural hazards agency should evaluate these measures on a continuing basis, recommending new policies as necessary.

Educational programs on seismic safety are essential to train professionals and other workers, and should be promoted by state governments. State seismic safety or natural hazards agencies should review pre-entry training and professional development programs, emphasizing awareness of seismic safety needs, and furthering high levels of competence in preparing earthquake-resistant designs. A state seismic safety or natural hazards agency should regularly review the performance of such educational programs, recommending new or improved programs as necessary.

Dealing with Critical Facilities

Critical facilities are those that house or serve large numbers of people, or otherwise pose unusually high hazards in case of damage or malfunction due to earthquakes. Facilities potentially highly hazardous include dams and reservoirs, nuclear reactors, liquefied natural gas (LNG) facilities, structures containing large quantities of explosive or toxic materials, buildings that are tall or house large populations, (e.g., high-rises, schools and prisons), and *types* of buildings that are of major concern because they are numerous (e.g., old hazardous buildings). Other critical facilities whose failure to function would cause grave concern are: hospitals; fire, police and emergency service facilities; and utility "lifeline" facilities, such as water, electricity and gas supply, sewage disposal, and communication and transportation facilities.

The magnitude of the potential hazard to critical facilities means that the state or federal governments must take special responsibilities for their safety. Appropriate measures include adequate processes for reviewing their safety, as well as other standards, policies and programs, outlined below, aimed at insuring satisfactory performance during earthquakes.

Admittedly it will be much easier to deal with the safety of new facilities when in the design stage than to strengthen or repair existing facilities. Some of the latter may be comparatively inaccessible or have hazards that will be costly to correct. Nevertheless corrective programs may be able to handle the most serious hazards in a reasonable time. Moreover the cost of making existing facilities safe emphasizes the importance of insuring that *new* installations are designed for adequate resistance to potential earthquake shaking.

Review of Critical Facilities: General

New critical facilities should be built to the best current state-of-the-art standards of design, including seismic resistance considered suitable to the region and the type of structure. (See above, especially p. 13, for comment on hazard assessment for critical facilities.) To insure compliance, the design and construction of critical facilities should be subject to appropriate independent review processes, using qualified personnel capable of exercising independent judgments on the adequacy of facility designs and construction.

Personnel conducting reviews need to be protected from political pressures, and should also be free from conflicts of interest. These goals are not easy to achieve in controversial matters.

A state seismic safety or natural hazards agency should evaluate the adequacy of the safety regulations and review processes governing critical facilities. For good results, the agency should operate from a position of independence, and should not simply be one of the old-line departments given new duties.

A Cause for Special Concern: Facilities Lacking Adequate Controls

Many kinds of critical facilities appear to be under no effective regulatory control having adequate regard for their seismic safety. These include older hospitals, essential emergency service structures (e.g., fire and police), liquefied natural gas (LNG) and similar terminal and refining facilities, buildings or structures storing toxic or explosive materials, and certain tall buildings (including their electrical and mechanical facilities).

The seismic safety of these structures and facilities needs to be evaluated, using appropriate standards. They should not simply be required to meet conventional building code requirements. The latter are unlikely to provide the higher levels of safety appropriate for critical facilities, whose potential failure would be grounds for unusual concern.

Dams and Reservoirs

Dams and reservoirs can pose very large seismic hazards, and all should be subject to careful evaluations by independent, highly qualified review bodies. Since 1929, California has provided for an independent review of the design and construction of all non-federal dams above a designated size, and for regular inspection of such dams as long as they remain in use. California's program is discussed on pp. 51-54. Even with respect to federally owned dams, the State of California and the U.S. government may be moving toward a cooperative arrangement giving the state a de facto right to review and comment on the seismic and geologic safety of proposed federal dams and reservoirs. Cooperative review has taken place in the case of Auburn Dam, stimulated in part by the California Seismic Safety Commission's Auburn Dam resolution outlining criteria for effective review processes and policies that could be adapted to apply to all dams. (See "A Note Respecting Federal Dams," p. 53, and Appendix D: "Auburn Dam Seismic Safety Review," a policy statement adopted by the California Seismic Safety Commission, March 10, 1977.)

Transportation Facilities

All major modes of transportation–railroads, trucks and buses, ports and shipping, and airplanes–are vulnerable to interruption by earthquake and geologic hazards. While the availability of transportation services is a critical need after all major disasters, it is essential to emphasize the principle of redundancy, so that if one or more modes fail in a major earthquake, others may be kept in operation. Thus it is imperative that the seismic safety of *all* transport modes and facilities be reviewed, and appropriate remedial measures taken.

A state transportation agency should be responsible for the seismic safety of highway systems, and should give special attention to (1) the safety of long bridges and many older bridges; (2) the stability of approach structures and fills; and (3) the hazards posed by poor foundation soils. To obtain an impartial evaluation of probable performance, it may be necessary to provide for an independent review of such facilities.

The importance of air transport emphasizes the need for concern about the seismic resistance of airports and landing areas. Especially important are control tower safety and the reliability of radio communications, including the capacity to function during and immediately after earthquakes. The state transportation or aeronautics agency should adopt appropriate policies to this end, and the federal government also has a clear concern. The same goes for local governments, which construct and manage most of the nation's airports.

Utilities

The earthquake performance of the principal public utility systems–e.g., electrical, gas, water and sewer–are of major concern, because large-scale or lengthy service interruption in time of disaster would pose widespread hardships and dangers. Consequently the suppliers of utilities should develop disaster plans for limiting damage, providing at least basic lifeline levels of service during and after earthquakes, and promptly restoring full service.

Appropriate control over utilities should be exerted by several state agencies, including those responsible for (1) utility regulation, (2) public health or industrial safety, (3) highways, (4) energy, (5) water resources, and (6) fire safety. Presumably principal responsibility for reviewing disaster plans and enforcing seismic safety regulations would rest with the state agency regulating utilities. But the other agencies should be required to review seismic hazards of utilities in their areas of concern, to insure maximum practicable safety in future construction and operation, as well as in disaster planning. In addition, state seismic safety or natural hazards agencies should participate in these reviews to be satisfied of their adequacy. It should be anticipated that many of these matters are likely to be highly controversial.

High-Rise Structures

High-rise structures—defined here as those over eight stories tall—present many difficult earthquake-hazard problems. These include failure of the structures themselves; failure of mechanical or electrical facilities; failure of elevators, stairways, and other means of egress; and vulnerability to fires, plus the extreme difficulty of fire-fighting especially at the higher levels. Minimum safety needs include formulation of plans for fighting fires in high-rises, rescuing occupants stranded by elevator and stairway failure, evacuating occupants, if possible, when fires occur, or providing the best available temporary shelter in "safe internal environments" if evacuation is not feasible.

Old and Hazardous Structures

Large numbers of hazardous buildings, many but not all of older vintage, are found in most urban areas. They are subject to collapse or severe damage in case of earthquakes. This is especially true of those built with unreinforced masonry. A state seismic safety or natural hazards agency, plus other appropriate state agencies, should devise workable programs to strengthen or eliminate these buildings, according to a reasonable order of priorities based on degree of hazard. (See also "Rehabilitation and Redevelopment," p. 30 below.)

Special reduced code requirements emphasizing *life-safety* are probably essential to facilitate progress in strengthening and rehabilitating such hazardous buildings. It is often economically unrealistic to expect that old buildings undergoing renovation can be brought fully up to the current seismic safety standards that apply to new buildings. Accordingly, reduced standards specifically designed for the seismic repair of old buildings can be justified if the strengthened buildings will then provide reasonable protection against significant injuries or loss of life in earthquakes. Recognizing this, in 1978 the California Seismic Safety Commission recommended permissive legislation allowing local governments to institute hazardous-building abatement programs employing reduced standards, and such a law was passed in 1979.

Schools

All public and private schools should provide high levels of security against death or injury by withstanding one or more earthquakes without substantial structural damage, and without causing death or injury to occupants. A state seismic safety or natural hazards agency should examine criteria and processes for reviewing school building design and construction in the interest of earthquake safety. (See pp. 40-42 for a discussion of California's Field Act, one useful model for state regulations governing the construction of safe public schools.)

Hospitals and Emergency Service Facilities

Regulations for design and construction of hospitals should make them sufficiently earthquake-resistant to minimize service disruption, to the extent practical. State seismic safety or natural hazards agencies should examine criteria and processes for reviewing hospital design and construction in the interest of acceptable earthquake safety levels. The California Hospital Act, and subsequent experience with its administration, provides one useful model. (See pp. 43-45 for discussion.) Analogous seismic safety regulations could be applied to other emergency service facilities.

Essential services such as hospitals, ambulance, fire fighting, police and other emergency services are often housed in unsafe buildings. Many old hospitals are earthquake-prone, as well as some of the newer ones–including for example some in California designed or built before the state's 1972 hospital safety legislation went into effect. They constitute hazards that probably range from severe to extremely critical. Programs should be formulated for replacing or abandoning unsafe hospitals as soon as is practicable. The high cost of doing this will require careful consideration of priorities, focusing first on the most critically deficient structures. Meanwhile special remedial measures should improve the earthquake resistance of hospital buildings and their electrical and mechanical components. Emergency service buildings should also be strengthened or the services relocated.

Communications

When disasters strike, many people–especially commuters in automobiles or travellers away from home–turn first to radio for information.

Accordingly it is highly desirable that major radio stations be capable of functioning during and immediately after strong earthquake shaking. Station facilities should be reviewed to evaluate ability to meet such needs for earthquake disaster information. The safety of other communications facilities, including TV and the emergency telephone systems, should be reviewed for reasonable ability to resist earthquake effects, and to continue functioning or to resume operation as soon as possible after an earthquake.

Nuclear Facilities

The safety of nuclear power plants and other facilities using radioactive materials is the principal responsibility of the federal Nuclear Regulatory Agency. Nevertheless, as in the case of federal dams, state seismic safety or natural disaster agencies may need to review processes governing the *seismic* safety of nuclear installations.

Planning, Development and Land-Use Control

State Land-Use Policies

State, regional and local policies on the use and development of land– e.g., planning programs, zoning regulations and permit decisions– should take seismic, geologic and other natural hazards into consideration. Such actions or decisions should (1) avoid creating new hazards or increasing the danger to life or property, and (2) whenever possible, reduce the levels of existing hazards, thus helping mitigate future disasters.

State planning legislation should require that seismic safety needs be acknowledged, and insist that adequate earthquake-hazard reduction measures be included in the land-use and development plans of state, regional and local governments. Requirements that public agencies prepare general plans should call for inclusion of a seismic safety element as an integral part. An appropriate state agency, such as a state seismic safety, natural disaster, or planning agency should prepare model seismic safety plan elements to guide other governmental agencies in their work on such elements. (California's effort to require local seismic safety plan elements is discussed on pp. 48-50.) The elements should be reviewed and revised regularly, and each agency should also evaluate its implementation programs on a continuing basis.

The state should advise and provide financial assistance to local governments and other agencies that may lack the staff or expertise to prepare and implement seismic safety elements for their plans.

An appropriate state agency should regularly review the plans and monitor the performance of state, regional and local agencies, in order to advise administrators and legislators on the adequacy with which seismic safety needs are being met. Where necessary the safety agency should recommend new legislation, stronger enforcement, or other appropriate actions needed to insure acceptable seismic-safety levels and reasonable progress in earthquake-hazard reduction.

States known to have substantial hazards should require special studies with respect to permits for construction in or near hazardous areas. The California Alquist-Priolo Special Studies Zones Act of 1972 requires such studies in the vicinity of important earthquake faults delineated by the State Geologist and found to be "sufficiently active and well-defined as to constitute a potential hazard to structures from surface faulting or fault creep." (Discussed on pp. 46-47.)

Rehabilitation and Redevelopment

The inability of existing structures to meet acceptable levels of seismic safety, and their potential failure in an earthquake, should be important considerations in deciding on the rehabilitation or removal of such structures in urban areas.

Where hazard levels are significantly high, but early compliance with current seismic safety requirements is prohibitively expensive, the state seismic safety or natural hazards agency should formulate special standards and guidelines for mitigating building hazards. These standards should provide for structural improvements adequate to minimize death and injuries. Buildings repaired under such standards—which will usually be less demanding than code requirements applying to new buildings—should be reviewed periodically to avoid indefinite perpetuation of the hazards remaining.

Following major earthquakes or other natural disasters, reconstruction or redevelopment should avoid perpetuating hazardous situations, and should strive to reduce future threats to life and property. To this end, the state, regional and local governments should develop contingency plans and guidelines for post-disaster planning and reconstruction, and should implement and enforce them immediately after damaging earthquakes.

Major Subdivisions and Growth-Generating Improvements

Land-use changes that are likely to attract or serve many people, or to induce future growth (e.g., construction of large subdivisions, major developments and critical facilities) should not occur without adequate review and findings that the changes will not produce undue seismic and geologic hazards that might jeopardize public safety. Additional considerations include the probable cumulative impact of such developments, and the effect on future seismic and geologic safety.

Major public improvements should not be located in areas having significant seismic or geologic hazards, unless appropriate alternative sites are unavailable. In any event, stringent measures should be taken in design and construction to provide acceptable safety levels for people and property. If such improvements are likely to generate future growth, plans and other regulations of public agencies should protect the seismic safety of future developments.

Emergency Preparedness and Post-Disaster Recovery

Responsibilities for effective emergency responses to disasters are shared by all levels of government, as well as the non-profit and charitable organizations that also have roles in disaster relief. Partly because preparing for disasters or responding to them is outside the usual routine business of government, it has proven difficult to achieve adequate levels of preparedness, especially for disasters such as earthquakes that do not occur frequently. Consequently vigilance and continuing effort are essential to help counter the inertia that works against effective pre-disaster preparation.

Emergency Response Plans

Each state needs a well-staffed, competent disaster or emergency preparedness agency to be a focal point for emergency planning and disaster-related information. Such agencies should also serve as crucial links between the state and federal governments in all phases of emergency planning. In addition, the individual state agencies and departments should each prepare their own emergency preparedness plans.

Every state government, as well as its principal local governments (e.g., municipalities, counties, and special districts), should prepare and regularly update plans for emergency responses to disasters considered likely to strike somewhere in the state. In seismic regions these would include earthquake disasters. State seismic safety or natural disaster agencies should regularly review and comment on the adequacy of state and local plans, investigate and report on performance during actual disasters, and where appropriate revise criteria for effective emergency response plans.

The states should require local governments' general plans to provide for post-disaster conditions. That is, local plans should be required to include contingency measures to be taken in heavily damaged areas, or in areas where land uses ought to be changed if redeveloped after a disaster. State monitoring of local performance will be necessary in the interest of uniform and consistent compliance.

Each state should have a statewide emergency medical care plan fully developed–and regularly updated–capable of efficient operation in any part of the state after a disaster.

A statewide emergency back-up system for inspecting disaster-damaged buildings should be developed. Rosters of competent personnel should be prepared, and *kept up to date*. In many major disasters it may be essential to bring in assistance from beyond the boundaries of a single state. Accordingly appropriate interstate or federal-state arrangements should be devised.

State and federal disaster-assistance systems should be evaluated and improved regularly. State seismic safety or natural disaster agencies should participate in these evaluations, recommending new state policies and legislation where needed. State bodies should also review and comment on federal disaster policies and performance. (See Appendix C for a detailed listing of recommended emergency preparedness measures, based on the report of the Advisory Group on Disaster Preparedness, California Legislature, Joint Committee on Seismic Safety, *Meeting the Earthquake Challenge* (1974).)

Post-earthquake Financial Assistance

Earthquake insurance presently can meet only a small part of post-earthquake monetary needs, because few property owners now carry such coverage. Accordingly, the federal and state government should explore ways of achieving far broader earthquake insurance protection.

Federal post-earthquake loans are essential, and inadequacies and inconsistencies in the existing disaster relief loan programs should be eliminated.

Even with substantial improvement, neither insurance nor disaster loans can satisfy all the post-earthquake need for financial help. Accordingly other sources of funds to help with recovery should be explored.

Post-earthquake Measures Immediately After a Disaster

State seismic safety or natural disaster agencies should prepare model local plans for post-disaster measures–such as immediate emergency surveys of building damage–as well as more detailed resurveys to guide decisions on whether to demolish or reconstruct seriously damaged structures.

States should establish central files consolidating all earthquake damage information. Seismic safety or natural disaster agencies should assist in this work.

Each state should provide for a system of reconstruction review, to develop plans and guidelines for the restoration of facilities and services in local communities after a major disaster.

Appropriate state agencies should prepare guidelines and perhaps model ordinances for post-disaster repair and reconstruction of damaged buildings. This will be a difficult task because of the differences among buildings and structural systems, and the variety of kinds of disaster damage that can occur.

Involving and Informing the Public

Information and education programs on disaster preparedness should be developed (1) to better inform the citizenry about potential hazards, including feasible precautions that can be taken and appropriate behavior in a disaster, (2) to further public understanding of the need to support disaster preparedness programs, and (3) to provide accurate public information about state and federal emergency preparedness plans, in order to obtain maximum public cooperation when disasters do strike.

Research Needs for Earthquake Policy Formulation

Many kinds of research studies would be helpful—and several of them essential—in supporting effective seismic safety policies. Especially important are better identification and understanding of geologic hazards; assessment of building performance in earthquakes; improved measures for saving lives; studies of social, economic and fiscal impacts; and development of appropriate responses to earthquake predictions.

Understanding Geologic Hazards

Further research is needed on the fundamental mechanisms of earthquakes, and their probable locations, frequencies and magnitudes, as well as the nature of associated ground responses. Such studies should include investigations of plate tectonics and its causes, the nature and processes of deformation of the earth's crust, regional and local strain accumulation, and mechanisms of crustal failure. Studies of individual major earthquake faults should explore their respective histories in order to better anticipate future earthquakes. Research on earthquake probabilities should be intensified and extended, as well as investigations of catastrophic phenomena associated with earthquakes, including soil failure, liquefaction, landslides, and tsunamis (seismic sea waves, sometimes called "tidal waves").

Dealing with Earthquake Predictions

The science of earthquake prediction, while now rudimentary, may within the next two decades permit forecasting the location and perhaps magnitude of at least some significant earthquakes. It is nevertheless likely to be a long time before most earthquakes can be anticipated accurately, and some may never be predicted precisely. This lends force to the California Seismic Safety

Commission's policy on earthquake prediction, which emphasizes the importance of a safer working and living environment as offering the best protection against casualties and damage in future earthquakes, with or without the capability of earthquake prediction.

In any event, in seismic regions society is entering a difficult "experimental" period. Although predictions may be unreliable, we cannot afford to ignore those that have some scientific legitimacy. Consequently government should develop processes for monitoring and evaluating earthquake predictions, and responding to the more credible ones. In this connection, the California Office of Emergency Services (OES) established the Advisory Group on Earthquake Prediction in March 1974–renamed California Earthquake Prediction Evaluation Council in April 1976–to prepare formal evaluations of those predictions deemed worthy of testing, and to prepare contingency plans. The council reports to the Director of OES, assessing the probability that a predicted event will indeed occur.[7]

Investigating and Mitigating Structural Hazards

Other research needs relate to the safety of present and future structures. The physical conditions of existing buildings and other structures should be studied to determine their resistance or vulnerability, estimate the damage they are likely to sustain, and evaluate the potential threat to life from their continued use. Applied research should investigate design criteria and performance of structures, with special attention to the implementation, enforcement expense, and cost-effectiveness of safety measures and designs. These studies should recommend ways to enforce current codes and insure that hazard-resisting design features are incorporated into buildings during construction and maintained throughout the useful lives of the structures.

Earthquake-resistant design criteria should guide engineering and economic decisions with respect to replacing, repairing, bracing, or rehabilitating existing buildings. Knowledge and expertise in improving the earthquake resistance of existing buildings should be collected, evaluated, and made available to owners and professionals. Engineering and structural safety criteria should be interpreted in terms of their economic and fiscal impacts.

Research on earthquake probabilities and hazard-mapping, including investigations of the causes of ground failure and the performance of soils during earthquakes, will facilitate better engineering and land-use decisions on the design and location of structures.

Reliable instruments maintained in good working order should be installed to provide "strong-motion" data on the kinds of shaking experienced by structures in future earthquakes. The information obtained will lead to better understanding of the kinds of local strong motions that earthquakes produce, as well as how the different kinds of structures respond to such motions.*

Learning to Save Lives and Minimize Injuries

Further research should estimate probable future life loss and casualties from earthquakes of different magnitudes and severity in various locales, and evaluate implications. The effects of loss of key services should be examined, along with the consequences of failures of critical facilities, or others that may endanger large numbers of people.

The effectiveness of response to earthquakes should be studied, along with emergency procedures for other kinds of disasters. Such investigations can guide the design of improved disaster preparedness measures.

Methods should be developed, through case-study investigations and other research, to evaluate the state of readiness of stand-by facilities and personnel at local, state and federal levels. These should include analyses of emergency procedures, and also provide for both scheduled and unannounced trial runs of emergency response plans. Private-sector response-readiness should also be studied, especially in the case of utilities and other major services.

The effectiveness of responses to disasters of many kinds, including earthquakes, tsunamis (seismic sea waves), hurricanes, tornadoes, floods, major explosions, and volcanic eruptions, should be evaluated. If any such disasters have been reliably predicted, governmental response and public experience should be examined to determine (1) problems encountered in dealing with the predictions, (2) modifications in responses after successive experiences with predictions and disasters, and (3) significant problems confronting pre-disaster preparation and post-disaster recovery. Despite the great differences among types of disasters, through careful comparative studies and drawing analogies it may be possible to learn a good deal from responses to non-earthquake disasters.

---

*In 1971 California established a program to install and maintain strong-motion instruments in representative geologic environments and structures. (California Statutes, 1971, ch. 1152.) The program is financed by small fees charged to applicants for city or county construction permits. By 1979 the program had reached "minimum statewide coverage." Field installations are about 50 percent complete. Building instrumentation is about nine percent complete, and will continue at the present rate of about ten buildings per year. Tom M. Wooton, "Strong Motion Instrumentation Program," *California Geology* 32 (4): 77-79 (April 1979).

Assessing Social, Economic and Fiscal Effects: Costs of Earthquakes and Seismic Safety Measures

Improved knowledge about community hazards and risks can guide local planners in addressing both the pre-and post-earthquake periods, and promoting better integration of post-earthquake recovery efforts into community planning processes. Such measures can capitalize on the opportunities to use community planning techniques in furthering seismic safety.

Estimates of the social, economic and fiscal effects of earthquakes, and of remedial programs, should be developed. As noted earlier, many hazardous buildings in urban areas are occupied by lower-income populations, minorities, the elderly and businesses that may depend on low rents. The socioeconomic effects of severe earthquakes should be considered in light of these facts.

To guide research into fiscal impacts, potential damage for projected earthquakes of varying severity should be estimated. The costs should be presented in both economic and non-economic terms, including lives possibly lost and injuries sustained, value of property lost or damaged, cost of recovery and rehabilitation, and losses due to adverse effects on the economy.

If old buildings are renovated and rents raised, the effect on the former residents and businesses must be considered in weighing the desirability of providing public financial assistance to reduce hardship.

Formulas for sharing losses and distributing costs of pre-earthquake preventive and remedial measures, and of post-earthquake rehabilitation and recovery programs, should be studied. The advantages, disadvantages and other implications of various formulas for cost-sharing and governmental subsidy should be examined.

After community leaders have reviewed preliminary findings with respect to cost-sharing formulas, further inquiries could determine public attitudes toward ways of paying for earthquake safety and providing compensation for earthquake damage.

Statewide tax and economic studies should help formulate "carrot and stick" policies to encourage and partially subsidize owners' efforts to improve the safety of hazardous structures, as well as decisions to reduce the occupancy of hazardous buildings or to abandon or demolish them altogether. Such aids should be accompanied by measures that impose some sanctions or withdrawal of benefits from owners who do not respond.

Some important rehabilitation programs can be scheduled in connection with major public works enterprises. Where abandonment, demolition and relocation are essential, ways can be sought to encourage and guide the transition, as well as cushion the shock.

Responding to Earthquake Research: Informational and Educational Follow-Up

In rapidly advancing fields of inquiry, those who depend on technical and scientific knowledge for their professional practice, or to guide them in discharging public responsibilities, need to have available the latest and best information in clear and usable form. Informational and educational efforts to achieve this goal are badly needed in many fields, including seismic safety.

Moreover scientific and technical knowledge has to be translated into public policies, guided by a sound understanding of the alternative choices facing society, and the costs and benefits of each. For this, organized study groups should examine earthquake research findings, especially those concerning risk, hazard-mitigation, casualties, losses and economic and fiscal impacts. Such investigations should evaluate remedial measures with regard to their necessity, equity, feasibility and acceptability. These studies can guide the formulation of priorities for disaster preparedness and response.

Promoting Public Support for Seismic Safety

State seismic safety or natural disaster agencies should prepare periodic well-written and informative status reports on seismic safety, indicating the degree of progress or lack of it. Pamphlets, films, newsletters, and other information media should be distributed, emphasizing preventive seismic safety measures. Material on earthquake hazards and seismic safety should be made available to schools at all levels. Scientific and technical advisors and skilled science writers should assist the news media in preparing and disseminating accurate information about earthquake hazards and seismic safety.

Information for Personnel with Special Functions

Strenuous efforts should be made to reach all practicing professionals with these attempts at information transmission and improvement of expertise. Others should also be involved, in addition to the earthquake professionals—e.g., taxpayers, property owners, administrators and policymakers, teachers and educators—in stimulating interest in and awareness of seismic hazards and what can be done about them.

Accordingly, courses, seminars, workshops and conferences should be directed to audiences of persons who have special roles in seismic safety activities, such as civic leaders and public officials; building inspectors and plan checkers; engineers, architects and planners; and real estate professionals and entrepreneurs (lenders, developers, brokers, salesman). These efforts should be repeated regularly in order to reach new personnel moving up into positions of responsibility, as well as to renew the interest and refresh the memories of others.

Information About Seismic and Geologic Hazards

Statewide clearinghouses should assemble information from the geologic and seismic literature and other relevant sources, and make it readily available for potential users, including the public.

Geologic mapping should show seismic hazards, and other pertinent geologic data should be provided to inform the general public, local governments, and other interested parties.

Laws or regulations governing practices in the real estate market should require comments of a geologic nature to be included in all public reports on subdivisions, thus providing vital background information to potential buyers. In the interest of such public disclosure, a 1978 California law requires any soils or geologic reports prepared specifically for a subdivision to be noted on the final subdivision map, along with the sources of the reports.[8] Such reports must also be referred to when submitting to the Real Estate Commissioner a comprehensive statement of information required for any subdivided lands proposed to be sold. The appropriate city or county government must keep copies of these reports on file for public inspection.

Prospective buyers should be informed about the nature of seismic hazards, in order that they may have seismic safety inspections made by competent engineers and other professionals. Building departments and private-sector professionals should be prepared to supply geologic data, as well as information on the seismic safety of buildings they have inspected. Where necessary, building departments should be ready to make inspections to assist prospective buyers, tenants and other interested parties, the costs being charged to those requesting inspections.

Information to Guide Individual Behavior

Information should be made available to the public concerning: (1) where to find help in evaluating hazards to homes or other property, (2) how to get help in improving the safety of such property and (3) what to do in emergencies. Other means of informing the public include earthquake exercises at schools and offices, use of audiovisual media, and printing emergency instructions in widely used documents like telephone directories. (Most telephone directories in California now include such a "survival guide.") Information on recovery aid should be distributed immediately after a damaging earthquake.

Seismic Safety of Public Schools: California's Experience

California's Field Act, passed in 1933, applied to public schools constructed after its enactment. Subsequent legislation and Attorney General's opinions attempted to deal with "pre-Field Act" buildings posing hazards.

The Field Act: Insuring the Safety of New Schools

The Field Act was enacted as a prompt response to the 1933 earthquake that destroyed or seriously damaged many public school buildings in the Long Beach and Los Angeles areas.[9] The law's purpose was to ensure that future public school buildings would be designed and constructed with sufficient earthquake resistance to protect occupants from death or injury.

The act provides for state supervision of the construction of public school buildings used for elementary, secondary, or community college purposes, or for the alteration or reconstruction of any such school if the estimated cost exceeds $20,000.[10] One of the principal provisions requires that all construction plans be prepared by qualified persons and that the designs be checked by an independent state agency. The independent review is generally considered most important, because it catches design errors, omissions or other inadequacies before construction contracts are let. The Structural Safety Section of the Office of the State Architect, in the state's Department of General Services, administers the act, which includes provision for checking plans and providing construction supervision. The program is financed from fees paid by school districts, based on construction costs.

Another very important provision of the act requires observation of the construction by the responsible architect or engineer, and continuous inspection of construction by a qualified person retained by the school board, to see that all requirements of the approved document are carried out. Moreover, all private-sector parties involved in the construction supervision, including the architect, engineer, inspector and contractor, must submit verified reports stating that the approved plans and specifications were complied with in construction. The state is also authorized and required to inspect the construction as needed to verify that all parties discharge their responsibilities under the Field Act, as well as to enforce the law.

The administering agency is authorized to adopt regulations stating the safety criteria and procedures, which are revised periodically to incorporate new knowledge gained from earthquake experience and research. Observers consider the Field Act highly successful in assuring reasonable compliance with acceptable standards of earthquake resistance. Since the law's passage, almost all schools built under Field Act provisions have performed well in earthquakes. While some experts anticipate that some Field Act buildings may experience substantial damage in future great earthquakes, there is agreement that the act's requirements have greatly reduced the potential for injury or life loss.

The Garrison and Greene Acts: Abating the Hazards of "Pre-Field" Schools

The Field Act's demonstrated success was not matched by equivalent progress in abating the hazards of "pre-Field Act" buildings. A decade ago California still had large numbers of hazardous pre-Field Act schools, whose renovation or replacement posed formidable problems because the state and its school districts were comparatively slow in dealing with their hazardous schools.

Much earlier, by the late 1930's, it had already become clear that because the act had no retroactive provisions, only limited progress was being made toward reducing the hazards of pre-Field Act schools. In 1939 the Legislature passed the Garrison Act to help meet the problem by outlining corrective steps, including having buildings inspected by qualified personnel, and financial procedures to follow if a building were found unsafe.[11] While some school districts made good headway in their hazards-reduction programs, many others largely ignored both the hazards and the Garrison Act.

The Greene Acts (three amendments to the Garrison Act), plus a state Attorney General's opinion, changed this pattern of nonresponse. A 1966 opinion based in part on 1963 tort liability legislation, and reaffirming a related 1964 opinion, found that school board members could be held personally liable if they failed to take due care in the interest of earthquake safety and consequently allowed the use of buildings that did not meet safety standards.[12]

The first Greene Act, passed in 1967, required a structural examination by January 1, 1970 of any public school building not constructed (or reconstructed) under the Field Act, and a report to the state as to whether the building had been found to be "safe" or "unsafe" under specified safety criteria.[13]

School board members were given immunity to personal liability if they initiated action to have non-Field Act schools examined, and when found hazardous, either took remedial action or posed the issue to the district voters. A second Greene Act in 1968 required that any school building so examined and found "unsafe" not be used for school purposes after June 30, 1975, unless repaired.[14] A 1974 Greene Act permitted a two-year extension of the 1975 deadline if a contract had been let and repair work started before June 30, 1975.[15] In addition to the legal actions noted, massive state funding of school building construction—including some bond financing provided expressly for the "unsafe" school problem—was an important factor in achieving earthquake safety in California schools.

In short, progress resulted from the combined effects of state financial assistance, assigning school board members personal liability for failure to exercise due care, a deadline for safety inspections, and the forced abandonment of unsafe buildings. From the original passage of the Field Act, the entire process of identifying and repairing or retiring unsafe public schools took more than 40 years.

Seismic Safety of Hospitals: The California Law

The principal goal of California's Hospital Act of 1972 was to insure, insofar as practicable, that all new hospitals would be designed and constructed to standards enabling them to remain in operation for emergency and other hospital services during and immediately after an earthquake.[16] The 1972 law used the Field Act as a guide, and required that design work for new hospitals or substantial additions or alterations to existing hospitals be done by qualified specialists, e.g., architects, structural engineers, and engineering geologists. The design must then be reviewed for structural safety by the Structural Safety Section, Office of the State Architect, under contract with the Office of Statewide Health Planning and Development, which is the administering agency. Also under contract, the Division of Mines and Geology reviews the technical adequacy of geological and seismological reports that, unless waived, are required to accompany hospital building proposals. A building safety board in the Office of Statewide Health Planning and Development acts as an appeals body for questions relating to the seismic safety of hospitals.

Although the law is considered successful, there have been problems with it, some of which have already been corrected by subsequent legislation. Thus a 1976 amendment redefined the term "hospital building" to exclude buildings with only outpatient services, and single-story, wood-frame buildings used only for skilled nursing or intermediate care facilities.[17] These exclusions were introduced partly because such facilities do not provide the kind of general hospital services required to meet disaster emergencies, as well as because single-story, wood-frame buildings are recognized as normally being relatively safe during earthquakes.

Other problems include the definition of and provision of "essential services." A recent task group report notes that the act "takes care of the building elements of the hospital but not the contents. . ."[18] A 1978 law goes some way toward meeting this deficiency by requiring the state to set standards for fixed hospital equipment anchorages, and calling for state approval of such anchorages before installation.[19] These standards relate only to anchorages, not to the strength of the equipment as such. Thus until equipment is also required to be earthquake-resistant, we may have the anomaly of seismically-resistant anchorages firmly securing equipment that nevertheless becomes unusable after an earthquake. Moreover, still not covered are such things as the pharmacy and blood bank with their sometimes fragile contents, or such problems as access to hospitals, or dependence on vulnerable outside utilities for water, sewage, gas and electricity.

In major crises, authorities need to consider the availability of all hospitals in an entire region, plus public schools and other public buildings that can house hospital-related emergency services. Such matters are not dealt with under a law that applies only to individual hospital structures, considered singly and separately.[20]

Other problems include the fee schedules, which are based on average construction costs. Thus some larger hospital projects probably pay more than their proportion of the actual state costs for the act's enforcement, while some smaller jobs no doubt do not pay their full share. Also, some ambiguities have been noted respecting jurisdiction and duplication of state and local fees and inspections, as well as some significant delays in project approval, especially in the early period of the act's administration.[21]

Solutions to the problem of undue delays may include experimental "fast tracking" methods of incremental project approval. Presumably the jurisdictional uncertainties have been cleared up by a recent opinion of the Attorney General holding that state review and approval of health facility design and construction have wholly preempted local action in that field.[22] According to this opinion, local jurisdictions may enact more restrictive construction standards relating to the structural safety of such facilities, but *enforcement* of such standards has been preempted by the state.

The earthquake forces for which hospitals are required to be designed may be excessively high under current interpretations,[23] and it has been estimated that future new hospital construction costs may be increased up to eight to ten percent by the 1972 Act, and probably somewhat higher. So far, however, there has been little experience with new hospital construction under the act, as most of the work so far reviewed has been limited to alterations, modifications and additions to existing hospitals.[24]

Some engineers maintain that the Office of the State Architect may have proceeded with undue dispatch in formulating and approving the regulations implementing the hospital act. A consultant's recommendations were used, and some of the typical processes in the public review of draft regulations by professional societies and other interested parties were telescoped in the interest of expeditious action. While critics acknowledge that review of draft regulations can take a lot of time, especially if allowed to drag on at length, they nevertheless believe that such reviews comprise an important phase of "due process," and should be included when future technical regulations are written concerning hospital safety or other aspects of seismic safety.

Like the Field Act, the Hospital Act applies only to *new* construction. Many older hospitals are probably in hazardous buildings, as well as some of the newer ones designed or built before passage of the Hospital Act. Many of these would probably become partly or wholly nonfunctional in a major earthquake.

In short, the Hospital Act represents a major step toward the seismic safety of an important type of critical facility. Its implementation process has turned up a number of problems that are being worked out administratively and through legislation. Undoubtedly the law will continue to be amended and reinterpreted.

Structural Safety in Hazardous Fault Zones: The California Law

In 1972 California passed the Alquist-Priolo Special Studies Zones Act, intended to regulate building and related activities in designated active earthquake fault zones.[25] The basic aim is to regulate development so as to guard against building activity—especially homes in subdivisions, but also other structures for human habitation—across active faults that could tear the structures apart or undermine their foundations and footings by surface faulting or fault creep. The act named the four most active known faults, and permitted the State Geologist to designate others.[26] Fault traces and the special studies zones encompassing them are delineated on 1:24,000 scale maps, preliminary maps are sent to the affected cities and counties for review, and official maps are then provided to each city and county having jurisdiction over lands within the designated regulatory zones.

The local government must review and approve geologic reports provided by applicants for all defined projects located within a zone, under policies and criteria outlined by the State Mining and Geology Board. The act applies to all land divisions on which structures for human occupancy are planned, as well as to all structures, per se, intended for human occupancy, *except* for single-family, wood-frame dwellings not exceeding two stories, that are planned as part of a development having no more than three such dwellings.

The administrative criteria adopted by the State Mining and Geology Board provide that no structure for human occupancy may be placed across the trace of an active fault.[27] Moreover, the area within 50 feet of an active fault is assumed to be underlain by active branches, unless geologic investigation proves otherwise. In any event, an application for a development permit for any project within a designated zone must be accompanied by a geologic report prepared by a registered geologist, unless waived by the local government under a procedure requiring a "no-hazard" finding, and with the approval of the State Geologist. A registered geologist retained by the local government must evaluate the reports.

When first administering the act, the Division of Mines and Geology used policies that were broader and more inclusive than those now employed.[28] Initially, in addition to including the major active faults on the map of special studies zones, the division also considered mapping "potentially active" faults (defined as those having evidence of surface movement during Quaternary time, i.e., in the past two or three million years) but found this policy impractical as it would have caused hundreds of "potentially active" faults to be included. Consequently, the division now maps only those faults believed to have a "high potential for surface ground rupture," as evidenced by surface displacement in Holocene time, i.e., the past 11,000 years.[29] Moreover, under the new policy a fault must be "well-defined" to be included, which means that at least some part of the fault can be seen at the surface and thus located with reasonable precision and confidence.

A 1977 study by Seismic Safety Commission staff found a good deal of local government uncertainty and confusion as to how the act and criteria should be applied. The treatment of single-family dwellings was perhaps the most common misunderstanding.[30] The act exempts certain single-family dwellings from geologic report requirements, but the criteria implied that no structures for human occupancy could be built on active faults. Accordingly some local governments interpreted the law and criteria as an *absolute* prohibition against construction on active faults, thus in effect including all single-family dwellings, although as noted the act intended that certain dwellings (up to three units) be excluded.[31] This confusion was removed with the adoption of revised criteria in January 1979, clearly exempting single-family dwellings individually or in groups up to three dwellings.

The Seismic Safety Commission staff study—based largely on a questionnaire survey of affected cities and counties—disclosed several other problems of ambiguity, definition and interpretation of the kind that should be expected when state and local agencies must implement a new law dealing with complex and unfamiliar matters. The staff report suggested a number of changes in the law—and in the policies and criteria established by the State Mining and Geology Board—to resolve questions of definition and application, and recommended preparation of a handbook to make the law more understandable.[32] These changes have in large part been approved by the Seismic Safety Commission and the Mining and Geology Board.

City and County Plans: California's Mandated Seismic Safety Element

Shortly after the San Fernando earthquake of 1971, California's basic urban planning legislation was amended to require that each of the state's cities and counties adopt a seismic safety element as part of its general plan.[33] In 1973 the California Council on Intergovernmental Relations issued advisory guidelines for such plans, including the seismic safety elements.[34] Among other things, the guidelines emphasize the importance of the relationship between the seismic safety element and other local plan elements for which it should provide essential primary inputs with respect to land use, housing, open space, circulation, and safety. In the council's words:

> The effect of . . . [the law] is to require cities and counties to take seismic hazards into account in their planning programs. All seismic hazards need to be considered, even though only ground and water effects are given as specific examples. The basic objective is to reduce loss of life, injuries, damage to property, and economic and social dislocations resulting from future earthquakes.[35]

A 1975-76 review by a committee of the Seismic Safety Commission found a good deal of progress, but also many problems and evidence of procrastination. Thus at that time 81 cities (of more than 400) and 19 counties (of 58) still did not have a seismic safety element in their general plans. Three years later, at the end of 1978, the commission reported 51 counties as having completed elements and the other seven as having them "in progress." For the cities, the commission reported that 45 had elements "in progress," seven still had no element, and all the rest had completed elements.

The law includes no provision for enforcement, nor is there any requirement for state review and evaluation of the local plan elements—except for a special review program for open space elements in connection with state open space funding. The commission's review committee itself evaluated a small but probably representative sample of seismic safety elements and found them highly varied.[36] Nevertheless, the requirement has moved local governments to identify seismic problems, and in some cases is "leading toward significant impacts on land-use decisions."[37] On the other hand, some governments are giving the requirement only token consideration or none at all.

With respect to background reports on which the elements were based, "[The]. . . review disclosed inadequacies, omissions, misinformation and inaccurate data."[38] The committee called for more extensive involvement of experienced professionals, and improved guidelines to help local governments do a better job. In 1978-79 new guidelines were prepared by the State Office of Planning and Research for use by local governments by the latter part of 1979.

In addition, a 1978 law now requires local governments to submit seismic safety elements and technical studies done in support of the elements to the state Division of Mines and Geology. At this time it appears likely that the elements and studies will only be catalogued and shelved. On the other hand, the file will be a resource for potential use by the division or by the Seismic Safety Commission in monitoring performance with a view to recommending future changes in the legislation.

Based on experience so far, the committee concluded that the requirements had brought "very significant benefits," commenting that most cities and counties had little awareness of earthquake problems before they went to work on their seismic safety elements.[39] Now, however, the topic is at least recognized by local governments throughout the state, and seismic safety policies are included in many of the local plans. On the other hand, preliminary studies by seismic policy researchers at the University of California, Santa Barbara, suggest that the local seismic safety elements have not so far proven particularly influential in altering policy in most jurisdictions.[40]

In short, the progress achieved is only a first step toward land-use measures that deal effectively with earthquake problems. The committee recommended a variety of state-level actions, mostly advisory or helping, rather than "get-tough" state enforcement measures. Further, the state should identify hazards of special state-level significance–perhaps developing appropriate hazards maps and other guides–and requiring local governments to address these in their plans. Finally, the committee urged that the effectiveness and longer-term impact of the local seismic safety elements be reviewed again in "several years."[41]

A 1977 survey of cities and counties in the San Francisco Bay Area by the Association of Bay Area Governments (ABAG) generally supported the findings of the seismic safety review committee. The ABAG report concluded that local governments' geologic and hydrologic policies are far more comprehensive than expected, but that only 70 percent of the region's local jurisdictions had adopted a seismic safety element or its equivalent, as of January 1977. With respect to the quality of the elements that were adopted, the ABAG study commented:

> The policies in most seismic safety elements are impossible to evaluate by themselves. If they are interpreted stringently, proposed development could be highly restricted. If not, they need not affect development.[42]

Finally, noting the fact that many engineers and geologists have been quite critical of much local work on seismic safety elements to date, the chairman of the Seismic Safety Commission's review committee—a planner by profession—countered with these comments:

> I sometimes wonder whether the engineers and geologists appreciate the newness of seismic concerns to local government, or the pressure of the many competing concerns, or the financial or other limitations under which such governments labor. Also I wonder whether they appreciate the need for a broad policy adopted and accepted long before detailed regulations and engineering studies are adopted and implemented. (Letter from George Mader, January 8, 1979)

In short, he believes that substantial progress has been made in alerting local government of the need to consider seismic safety, while acknowledging that we still have a long way to go.

California's Dam Safety Program

California's dam safety program was enacted in 1929, in response to the catastrophic 1928 failure of St. Francis Dam with heavy loss of life and major property damage. Investigations after the dam's collapse showed that careful studies of the project design and of the site geology should have disclosed the makings of the disaster. The following comment in the verdict of the Los Angeles County coroner's jury investigating the tragedy outlines the concept of independent review on which California's dam safety program is based:

> A sound policy of public safety . . . and engineering judgment demands that the construction and operation of a . . . dam should never be left to the sole judgment of one man, no matter how eminent, without check by independent expert authority, for no one is free from error, and checking by independent experts will eliminate the effect of human error and insure safety.[43]

In implementing this concept, the 1929 Dam Safety Act established the Division of Safety of Dams and made it responsible for evaluating plans and specifications for the construction of new dams, or for modifications of existing dams.[44] In addition, the division supervises construction and operation of dams, and maintains a continuing surveillance program, monitoring and inspecting dams and reservoirs annually, or more frequently when considered necessary.

All dams in the state above a minimum size are included, except for those owned and operated by agencies of the United States government (see note, page 53). Qualifying dams are those more than six feet high and with a storage capacity of more than 50 acre-feet, or dams more than 25 feet high and with a storage capacity of more than 15 acre-feet.

The division's jurisdiction begins when it receives a request to approve of the plans and specifications for a dam or reservoir. After careful study the application is approved if the division is satisfied with the safety of the plans, permitting construction to proceed. During construction, detailed inspections are made to be sure that the approved plans and specifications are followed, and that any new or changing conditions are satisfactorily taken care of.

When construction is complete, the division makes a final judgment of the project's safety, approving and certifying it for operation if the finding is favorable. The certificate may contain conditions of operation, and also is subject to amendment or revocation, after hearings, if later developments or new regulations raise substantial safety questions; public hearings must be held prior to a revocation.

After a project begins operation, the division initiates a process of inspection. Old dams as well as new ones are subject to this regular, continuing surveillance:

> Newer and more demanding safety criteria require new studies of old dams. These are expensive in themselves, and they have the potential of requiring expensive alterations. Structural modifications acceptable in the past may be inadequate for today and for the future. Performance standards change. The public will not always let us be complacent, and the dam engineer must not let past decisions stand unchallenged forever . . . .[45]

The program must be judged a substantial success. For one thing, dams and reservoirs subject to state review and surveillance have an excellent performance record during the half-century since the Dam Safety Act was passed. Of course, this evidence is not definitive, as seemingly safe dams or reservoirs can contain undiscovered flaws and fail after many years of uneventful operation. Moreover there has not been a really great earthquake in California since 1906. Meanwhile, the past fifty years seems to have provided a rather good interim test for California's dam safety program.

Both the law and its administration have been up-dated and improved on the basis of experience and new information. For example, after the 1963 Baldwin Hills reservoir failure in Los Angeles, the law was amended in 1965 to include reservoirs as well as dams, and offstream dams and reservoirs were added, except those with the principal purpose of storing water for agricultural use. The Baldwin Hills failure was caused by ground movement and subsidence associated with pre-existing faults through the site. The Department of Water Resources in their report on the investigation of the Baldwin Hills reservoir failure stated:

> Sitting on the flank of the sensitive Newport-Inglewood fault system with its associated tectonic restlessness, at the

> rim of a rapidly depressing subsidence basin, on a foundation adversely influenced by water, this reservoir was called upon to do more than it was able . . . .[46]

This prompted a special inspection and safety evaluation program for all dams, with those in urban locations given priority. Particular attention focused on geologic structure beneath the dams, and the potential for subsidence.

Similarly, remedial actions were taken after the 1971 San Fernando earthquake caused a near-failure of lower San Fernando Dam and temporary evacuation of 80,000 people who lived downstream. A subsequent report on the San Fernando earthquake included a special section analyzing the performance of soils and dams, and recommending improved criteria, requirements and procedures, most of them for implementation by the Division of Safety of Dams.[47] A recent inquiry into the status of the recommendations drew this response from the chief of the division:

> most of the recommendations which apply to this Division have been or are being implemented on a priority basis. As a result of experience gained, the others may no longer be necessary.[48]

In summary, the basic dam safety legislation appears quite sound. In fact, the United States Committee on Large Dams of the International Commission on Large Dams acknowledged its debt to California when it issued the 1970 model law for dam safety, which follows California's statute very closely.[49]

A Note Respecting Federal Dams

Federally owned dams and reservoirs, like other federal facilities, are not constitutionally under California's dam safety program. Nevertheless recent administrative arrangements have permitted the state to play an important role in the comprehensive safety review of at least one proposed federally owned dam. Prompted in part by the Oroville earthquake of 1975, and the Teton Dam disaster of 1976, the state initiated requests to be given a role in reviewing the safety of Auburn Dam, which was designed by the U.S. Bureau of Reclamation to be a 685-foot-high, thin, double-curvature concrete arch structure, located on the American River near Auburn, California.

After a good deal of negotiation and letter writing, and a formal resolution by the California Seismic Safety Commission (see Appendix D), arrangements were agreed upon between the U.S. Department of Interior (and the Bureau of Reclamation) and the California Resources Agency. Under these arrangements, appropriate state agencies, principally the Division of Safety of Dams and Division of Mines and Geology, conducted on-site inspections of Auburn Dam and received copies of all materials produced by the Bureau of Reclamation's own special analysis of seismic and geologic safety questions affecting the proposed dam. Lying behind the agreement is the nearly explicit understanding, reiterated in public testimony by federal officials appearing before the Seismic Safety Commission, that construction of the dam will not proceed unless the State of California is satisfied that the dam as designed is reasonably secure from collapse due to earthquakes or foundation failure. It is hoped that this experience is setting a precedent for qualified state safety agencies, acting on the advice of independent review bodies, to have a role in decisions relating to the safety of major federal dams in their areas. The inclusion of qualified state agencies early in the review process for federal dams should reduce the time and effort needed to satisfy state experts and officials that the proposed structures are safe, and should enable the state to fulfill its appropriate roles as principal guardian of the health, welfare and safety of its citizens.

The Secretary of the Interior emphasized the value of the review process in a letter to the California Seismic Safety Commission: "Great care has been taken by Interior agencies, the State of California, and private consultants, so that a sharing of reports and studies takes place before I make a final, unified, and carefully considered decision sometime after the first of the year. I believe a public review process conducted in this manner will serve to enhance the confidence of Californians in the judgments made by Federal agencies." (Letter dated October 23, 1978, signed by Secretary Cecil D. Andrus.)

Subsequently an expert consulting board reporting to the California Department of Water Resources recommended a set of restrictive seismic criteria. These effectively forced abandonment of the double-curvature concrete arch dam design, prompting the exploration of other, more traditional designs—e.g., a thicker concrete gravity dam or a rock-fill dam—as well as perhaps other sites.

Three Other California Programs: Inundation Mapping, Evacuation Planning, and Freeway Retrofitting

In the past decade, prompted largely by what was learned from the 1971 San Fernando earthquake, California has instituted three other programs intended to forestall future casualties and damage. These programs include mapping areas that could be inundated by dam failure, drafting evacuation plans for the residents of these areas, and "retrofitting" freeway overpasses to improve their ability to survive earthquake shaking.

Inundation Mapping and Evacuation Planning

The near-failure of Lower San Fernando Dam in the 1971 earthquake and the evacuation of some 80,000 people located downstream emphasized the importance of accurate foreknowledge about areas likely to be flooded by dam or reservoir failure, as well as the value of advance planning for emergency evacuation of residents. Accordingly a 1972 statute inaugurated an inundation mapping program, requiring the owners of dams to prepare and file inundation maps if it has been determined that partial or total dam failure would result in death or personal injury.[50] After reviewing the inundation maps, the state Office of Emergency Services (OES) is required to designate areas within which partial or total failure would result in death or injury, whereupon the public safety agencies of local governments having territory in any such areas are required to adopt emergency procedures and plans for the evacuation and control of populations below the dams.

The statute initially required three maps for each dam, designating zones flooded by failure when water storage is at full capacity, at median levels, and at normally low levels. A 1973 amendment reduced the required filing to one map showing inundation areas if storage is at full capacity. A 1974 amendment permitted the Office of Emergency Services to waive the map-filing requirement if an onsite inspection determines that the inundation potential can be ascertained without a map, or where adequate evacuation procedures can be developed without a map.

OES has administered the program by means of a series of letters sent to dam owners, informing them of the statutory requirements. Policies embodied in these letters have been reworked into draft regulations proposed for inclusion in the California Administrative Code. The draft is receiving administrative review, after which it will go through a public hearing process before being incorporated in the code.

The *mapping portion* of the program appears to have been quite successful.[51] Ninety percent of the required maps have been submitted by owners of dams falling under state jurisdiction, and have been reviewed and approved by OES. The remaining maps are either not yet due (because dam construction is not complete) or are in the process of being reviewed by OES.

The *emergency evacuation planning* portion of the program has been much less successful. Each local government having territory in an inundation area is required to prepare an emergency evacuation plan for that area. As things stand now, 2,000 evacuation plans are required, but only about 20 percent of the total have been received and reviewed by OES.[52]

Some critics suggest that in the early days of the evacuation planning program OES did not put enough pressure on local governments to secure expeditious action. This may be changing, as OES notes that local plans began to arrive in significant numbers in 1978. It is now estimated that OES will review most if not all of the 2,000 evacuation plans in 1979 and 1980.

Transportation Lifelines and Freeway Retrofitting

During the 1971 San Fernando earthquake, two major freeway interchange bridges under construction in or near the fault zone partially collapsed, sustaining about $6.5 million in damage during the 12 seconds of shaking.[53] The interchanges had not yet been opened for traffic, and in any event the earthquake occurred in the early morning hours, otherwise the failures would have caused fatalities, injuries and general disruption.

Prior to 1971, very few bridges in the continental United States had sustained serious seismic damage. The San Fernando event highlighted a number of deficiencies in bridge design. Most important was the failure to tie different segments of the bridge superstructure together properly. The second most serious problem was inadequate reinforcement of many of the support columns, whose reinforcing steel ties did not provide adequate confinement of the concrete in the columns.

With respect to all new work, the California Department of Transportation took immediate steps to redesign structures, correcting the deficiencies. By 1971, of course, many thousands of miles of freeways had already been constructed in California, averaging one bridge per mile of freeway. The huge effort required to repair the existing bridges, correcting deficiencies and making them more resistant to earthquakes—called "retrofitting"—was initiated in 1972, and has been pursued ever since. So far, the retrofitting program has focused on installing restrainers for the hinges and bearings on bridges in and near the Los Angeles and San Francisco Bay areas, which are both densely populated and more seismically active than some other regions. By 1980, 650 highway bridges will have been retrofitted at an estimated cost of $15 million, and a program is in progress to survey the state's entire stock of 14,000 bridges in the Interstate and Primary Highway system to identify those that are seismically deficient.[54] The retrofit program's goal is:

> . . . to make all . . . structures seismically resistant to the extent that they may sustain damage, but not collapse completely. We would like them to be able to carry at least a limited amount of emergency traffic even though they may be damaged.[55]

In preparation for potential blockages or gaps in important lifeline routes, the Department of Transportation is identifying bypasses and alternate routes. This work, as done for the Los Angeles region, is described as follows:

> The plans consist of 3,000 detailed freeway bypasses in 125 Los Angeles basin communities on the 650-mile urban freeway system. The plans define the alternate routes as well as document and furnish identification and deployment of traffic, engineering and police personnel from each agency involved, so that everyone knows his role in the plan . . . . in a major earthquake, appropriate combinations will be used to cover . . . widespread areas, and reestablish the life-line network.[56]

Conclusion

This discussion has relied heavily on California's experience in responding to earthquake safety needs. While several of the most important state programs were selected for treatment in previous chapters, this does not comprise a comprehensive discussion of all California's efforts relating to seismic safety. A recent report by the Seismic Safety Commission identified 52 state agency programs and activities, with an expenditure of approximately $18,260,000 in 1978-79.[57] In reviewing these activities, the commission grouped them in these six major categories—the percent of state earthquake-related expenditures allocated to each being indicated in parentheses: disaster preparedness (1.9 percent), earthquake information (2.2 percent), scientific investigations (15.7 percent), existing hazards and their reduction (50.3 percent), new construction (29.1 percent), and policy (0.8 percent).

California has made important commitments and taken many initiatives respecting seismic safety, and has achieved substantial progress. In the last decade, especially, we have learned much more about earthquake ground motion and its consequences, and increasing attention has been given to problems of engineering design. Building codes have also been significantly improved and prevailing construction practices up-graded.

On the other hand, while these accomplishments and the large number of seismic safety programs and dollar expenditures may look impressive, knowledgeable observers and critics are convinced that a great deal remains to be done. Many major hazards persist—for example, old and unsafe buildings—whose mitigation or removal will take major commitments of time and effort before long-term solutions are achieved. Moreover, short-term efforts also need strengthening because present state and local abilities to respond to major disasters appear questionable, compared with the resources that a large earthquake will call for. In weighing both short- and long-term needs, the Seismic Safety Commission report comments on existing hazards and disaster preparedness:

> Given the great number of hazardous structures in use today . . . built prior to any consideration of lateral force requirements [to resist sideways shaking], any program of rehabilitation and strengthening . . . must be directed toward . . . long term solution[s]. There is a general consensus . . . that in the short term (within the next ten or fifteen years), disaster preparedness can provide the greatest degree of hazard mitigation in terms of lives saved . . . .[58]

Despite this, as noted earlier, less than 2.0 percent of the state's expenditures for seismic safety is going into disaster preparedness. Some observers see this as evidence that the state is giving short-term mitigation measures and disaster preparedness much too low a priority.

Heretofore California has tended to approach earthquake safety by bits and pieces, often through ad hoc actions after specific seismic events, attempting limited-purpose responses to some major findings of vulnerability, as revealed during an actual earthquake. A classic example was the 1933 demonstration that school buildings were extremely vulnerable, resulting in the Field Act aimed at enforcing safety standards for all public schools.

This process produced important seismic safety progress on a few fronts, while comparative neglect continued on others. Consequently the Seismic Safety Commission was created with the quest for *comprehensiveness* as one of its principal objectives. At the heart of the commission's role is setting realistic goals and priorities and seeking effective ways of implementing them. A crucial part of the task is identifying and filling major gaps in seismic safety programs. This calls for continuing and steady overview of the adequacy of all seismic safety efforts, and keeping a watchful eye for hazards that are not being dealt with.

In pursuit of its objective, in 1979 after a two-year effort the commission adopted a major statement of goals and policies,

> . . . [to] represent the Seismic Safety Commission's present consensus on the most important seismic safety goals for Californians. This statement of goals was prepared to focus attention on the fundamental issues, to communicate what the Commission believes ought to be accomplished over the long run, and to help the Commission decide on specific work projects to be undertaken.[59]

The commission is distributing the document widely to stimulate and encourage individuals and organizations to assist in appropriate ways with the quest for seismic safety. The methodical statement of goals will also be a yardstick for the commission and others to use in regularly assessing progress and deciding when new legislation or other measures are needed. The policy document thus forms the "base-line" of the commission's current thinking, to be regularly reviewed and re-thought as more is learned about earthquake hazards, and about practical and effective ways of implementing seismic safety programs.

In short, California now has some basic programs and policies that are helping make better use of what we already know. It also has the means to facilitate learning from future earthquakes and applying the new knowledge. This is important because careful study is showing the value of lessons that can be learned even in small earthquakes. For example, a recent damaging California earthquake occurred in Santa Barbara and vicinity, August 13, 1978. While it was of modest magnitude–5.6 on the Richter scale–it did a surprising amount of damage and prompted a good deal of thought about seismic safety policies. For one thing, large numbers of mobile homes were thrown off their rather flimsy foundations, causing considerable financial losses, highlighting the vulnerability of mobile residences with inadequate foundation systems to minor shaking. This is of some significance because in California and in much of the United States mobile homes represent an increasing share of the housing stock.

In other ways the results of the Santa Barbara earthquake were disturbing. Damage claims amounted to $7 million, and there were 65 injuries. Moreover, the hardest hit site, the University of California campus, did not have its normal contingent of 14,000 students in residence. Many more injuries would undoubtedly have occurred if large numbers of students had been on campus. Ten of the University's 50 large buildings were significantly damaged, and investigation suggested cases where the codes were not followed, or where errors in construction appear to have occurred.

University of California Seismologist Bruce A. Bolt advised the Seismic Safety Commission (of which he is a member) that experience from Santa Barbara's small earthquake "is worth much thought."

> A small 5.6 magnitude earthquake is able to produce that kind of destruction in a limited area. What is to be done to prepare for a great earthquake? First, we need a detailed re-examination of earthquake risk in California. We need to look at it in a broader framework than . . . used in the past. We must consider earthquakes in relation to the whole economic system. We must determine how codes are being applied, how communities are responding to suggestions to mitigate . . . earthquakes, and particularly we must assess the cost/benefits that can assure that the expenditures of public funds can be fully justified and clearly efficient if we are to bring the situation up to a steady state [of adequate preparedness] within the next few years.[60]

Fortunately there appears to be a good deal of actual or latent public support for such efforts. For example, a recent survey of the human response to the Southern California Uplift—a widely publicized bulge in the earth's surface centered near Palmdale—found it "reassuring to realize the extent of potential public support for hazard reduction programs," and further noted evidence of "widespread public understanding of the need to prepare for earthquakes through programs aimed at reducing the hazard of earthquakes as well as through improving emergency response capability."[61]

California is attempting to work toward more comprehensive approaches to seismic safety, capitalizing on the basic public support that exists, and tailoring programs to meet its need, building on past experience as well as on what can be learned in the future. Similarly, other states can review their seismic-geologic situations in developing appropriate responses. They should be able to learn from what California has done—or failed to do as the case may be—adapting some of this state's policies where appropriate, or devising others where needed, in hammering out programs to fit circumstances and needs. Finally, they should recognize that California, despite its acknowledged "leadership" in seismicity and susceptibility to earthquakes, by no means possesses all the answers, and indeed still has a long way to go.

Notes

1. Clarence R. Allen, "Geological Criteria for Evaluating Seismicity," C. Lomnitz and E. Rosenblueth, eds., *Seismic Risk and Engineering Decisions* (Amsterdam: Elsevier Scientific Publishing Co., 1976), p. 65.

2. B.F. Howell, Jr., "Average Regional Seismic Hazard Index (ARSHI) in the United States," Douglas E. Morgan et al., *Geology, Seismicity, and Environmental Impact*, Association of Engineering Geologists (Los Angeles: University Publishers, Special Publication, October 1973), pp. 277-285.

3. The author gratefully acknowledges the assistance of Bruce A. Bolt, University of California, and Richard H. Jahns, Stanford University, in formulating this evaluation.

4. Lloyd S. Cluff, "Geologic Perspectives on Earthquake Hazards and Dam Safety" (Woodward-Clyde Consultants, lecture notes for Seminar on New Perspectives on Safety of Dams, Stanford University, August 28-September 1, 1978), p. 6.

5. U.S., Army, Chief of Engineers, "Attenuation of High-Frequency Seismic Waves in the Central Mississippi Valley," Report 10, Miscellaneous Paper S-73-1, *State-of-the-Art for Assessing Earthquake Hazards in the United States*, Otto W. Nuttli and John J. Dwyer (July 1978), pp. 6, 64, 72.

6. U.S., Army . . . "Specifying Peak Motions for Design Earthquakes," Report 7, Miscellaneous Paper S-73-1, *State-of-the-Art for Assessing Earthquake Hazards in the United States*, Ellis L. Krinitzsky and Frank K. Chang (December 1977), p. 9.

7. California, Earthquake Prediction Evaluation Council, "Earthquake Prediction Evaluation Guidelines," *California Geology*, 30:158-160 (July 1977).

8. *Statutes of California*, 1978, ch. 521 (*California Business and Professions Code*, sec. 11010, and *California Government Code*, secs. 66434 and 66435.5).

9. *California Education Code*, secs. 39140-39156 (elementary and secondary schools); and secs. 81130-81146 (community colleges).

10. The key operative language in the Field Act, as amended, is this:

> The Department of General Services under the police power of the state shall supervise the design and construction of any school building or, if the estimated cost exceeds twenty thousand dollars ($20,000), the reconstruction or alteration of or addition to any school building, to ensure that plans and specifications comply with the rules and regulations adopted pursuant to this article and building regulations published in the State Building Standards Code, and to ensure that the work of construction has been performed in accordance with the approved plans and specifications, for the protection of life and property. (*California Education Code*, sec. 39140).

11. *California Education Code*, secs. 39212-39232.

12. *Opinions of the Attorney General of California*, vol. 47, opinion no. 65-324, May 4, 1966, p. 163.

13. *Statutes of California*, 1967, ch. 214 (*California Education Code*, secs. 39210 ff.).

14. *Statutes* . . . 1968, ch. 692 (*California Education Code*, sec. 39227).

15. *Statutes* . . . 1974, ch. 190 (*California Education Code*, sec. 39227).

16. *Statutes* . . . 1972, ch. 1130. The law's enacting clause reads as follows:

> It is the intent of the Legislature that hospitals, which house patients having less than the capacity of normally healthy persons to protect themselves, and which must be completely functional to perform all necessary services to the public after a disaster, shall be designed and constructed to resist, insofar as practicable, the forces generated by earthquakes, gravity, and winds. In order to accomplish this purpose the Legislature intends to establish proper building standards for earthquake resistance based upon current knowledge, and intends that procedures for the design and construction of hospitals be subjected to independent review . . . . It [the law] shall be administered by the State Department of Public Health, which shall contract for enforcement of such provisions with the Department of General Services which now successfully enforces the provisions of the "Field Act."

17. *Statutes* . . . 1976, ch. 177 (*California Health and Safety Code*, secs. 15001 and 15001.5).

18. California, Seismic Safety Commission, *Report of the Task Committee of the Seismic Safety Commission on the Hospital Act of 1972* (SSC 77-03, May 12, 1977), p. 5.

19. *Statutes of California*, 1978, ch. 835 (*California Health and Safety Code*, secs. 15009, 15011.1, 15021 and 15022).

20. In providing post-disaster emergency care, it is important to consider joint use of facilities that remain functional after a disaster:

> It may be necessary for a hospital to be completely functional where there is only one for a county or . . . only one in a large region. But if several hospitals are nearby, would they all comply with the intent of the law if they pool their resources and capabilities by concentrating on the safety of a particular function in each hospital? (California, Seismic Safety Commission, note 18 above, p. 6).

21. California, Seismic Safety Commission, see note 18, above, pp. 8-14.

22. *Opinions of the Attorney General of California*, vol. 61, opinion no. CV 77-222, August 3, 1978, p. 3.

23. California, Seismic Safety Commission, see note 18, above, pp. 18-20.

24. California, Seismic Safety Commission, see note 18, above, p. 24. One estimate noted "about a 25 percent increase in . . . [the cost of] structural components of a hospital project," which in turn accounts for 12 to 15 percent of all project costs for most new construction. Totaling up percentage cost increases for other components, the estimate concludes that "the total project cost increase brought on by the new regulations is approximately 8 to 10 percent."

These figures do not include other costs that are often overlooked, for example, a processing cost of 0.7 percent, increased engineering and inspection costs, the costs caused by delays in a period of inflation, and other "red tape" costs. (Private communication to Stanley Scott, December, 1978).

25. *Statutes of California*, 1972, ch. 1354 (*California Public Resources Code*, secs. 2621-2630).

26. *California Public Resources Code*, sec. 2622.

> . . . the State Geologist shall delineate . . . special studies zones to encompass all potentially and recently active traces of the San Andreas, Calaveras, Hayward, and San Jacinto Faults, and other such faults, or segments thereof, as he deems sufficiently active and well-defined as to constitute a potential hazard to structures from surface faulting or fault creep. Such special zones shall ordinarily be one quarter of a mile or less in width . . . .

27. California, Division of Mines and Geology, *Fault Hazard Zones in California*, Special Publication 42, Earl W. Hart (revised January 1977), pp. 20-21.

28. California, Seismic Safety Commission, *Report on Local Governmental Implementation of the Special Studies Zones Act* (July 1977), pp. 17-19.

29. *Fault Hazard Zones* . . . , see note 27, above, p. 7.

30. California, Seismic Safety Commission, see note 28, above, pp. 8-9.

31. California, Seismic Safety Commission, see note 28, above, p. 9.

32. California, Seismic Safety Commission, see note 28, above, pp. 43-44. See also staff memo from Peter Stromberg, Seismic Safety Commission, to Special Studies Zones Committee, September 8, 1977.

33. *California Government Code*, sec. 65302[f] defines the required element as follows:

> A seismic safety element consisting of an identification and appraisal of seismic hazards such as susceptibility to surface ruptures from faulting, to ground shaking, to ground failures, or the effects of seismically induced waves such as tsunamis and seiches.
>
> The seismic safety element shall also include an appraisal of mudslides, landslides, and slope stability as necessary geologic hazards that must be considered simultaneously with other hazards such as possible surface ruptures from faulting, ground shaking, ground failure and seismically induced waves.

34. California, Seismic Safety Commission, *A Review of the Seismic Safety Element Requirement in California*, SSC 77-01 (March 25, 1977), Appendix A: "Guidelines, September, 1973: Seismic Safety Element," California Council on Intergovernmental Relations.

35. See "Guidelines," note 34, above, p. 23.

36. California, Seismic Safety Commission, *A Review*. . ., see note 34, above, p. 8.

37. Loc. cit.

38. Loc. cit.

39. California, Seismic Safety Commission, *A Review* . . ., see note 34, above, p. 17.

40. Letter to Stanley Scott from Dean E. Mann and Alan J. Wyner, December 8, 1978.

41. California, Seismic Safety Commission, *A Review* . . ., see note 34, above, p. 23.

42. Association of Bay Area Governments, *San Francisco Bay Region: A Review of Local Regulations Related to Geologic and Hydrologic Hazards, Constraints and Resources* (May 1977), p. 2.

43. Quoted in Clifford J. Cortright, Chief, Design Branch, Division of Design and Construction, [California] Department of Water Resources, "Report on States' Experiences and Their Problems in the Inspection, Maintenance and Rehabilitation of Old Dams" (pre sented to the Engineering Foundation Conference, September 23-28, 1973, Pacific Grove, California), p. 2.

44. *California Water Code*, secs. 6000-6501.

45. Gordon W. Dukleth, Chief, Division of Design and Construction, [California] Department of Water Resources, "Effective State Dam Safety Supervision," (presented at Seminar on New Perspectives on the Safety of Dams, Stanford University, August 28, 1978), p. 6.

46. California, Department of Water Resources, *Investigation of Failure: Baldwin Hills Reservoir* (April 1964), p. 64.

47. H. Bolton Seed, "Dams and Soils," California, Joint Committee on Seismic Safety, Special Subcommittee, *San Fernando Earthquake of February 9, 1971 and Public Policy* (Sacramento: July 1972), pp. 14-15.

48. Letter to Stanley Scott from James J. Doody, Chief, Division of Safety of Dams, California, Department of Water Resources, November 21, 1978.

49. United States, Committee on Large Dams of the International Commission on Large Dams, *Model Law for State Supervision of Safety of Dams and Reservoirs* (New York: United States Committee on Large Dams [1970]), pp. I and II.

50. *Statutes of California*, 1972, ch. 780 (*California Government Code*, sec. 8589.5).

51. Letter to Stanley Scott from H.R. Pulley, Senior Planner, Office of Emergency Services, November 29, 1978.

52. See note 51, above.

53. Oris H. Degenkolb, *Retrofitting Techniques for Highway Bridges* [Sacramento: Department of Transportation] n.d., p. 1.

54. Degenkolb, note 53, above, *Bridge Earthquake Restrainer Program* [Sacramento: Department of Transportation] April 1978, table.

55. Guy D. Mancarti and Oris H. Degenkolb, *Retrofitting Structures to Increase Seismic Resistance of Highway Lifelines* [Sacramento: California Department of Transportation] n.d., p. 6.

56. Ibid., p. 2.

57. This figure excludes expenditures under the special one-time project for restoration of the state capitol building. The total cost of this five-year effort is now estimated at $61 million, of which about $18.5 will go for strengthening structural and ornamental elements to resist earthquakes. California Seismic Safety Commission, *Report on State Agency Programs for Seismic Safety* (April 30, 1979), pp. 5, 80.

58. Ibid., pp. 7-8.

59. California, Seismic Safety Commission, *Goals and Policies for Earthquake Safety in California* (1979), pp. ii-iii.

60. Bruce A. Bolt, Statement before California Seismic Safety Commission at its regular meeting of May 10, 1979.

61. Ralph H. Turner, et al., *Earthquake Threat: The Human Response in Southern California* (Los Angeles: Institute for Social Science Research, University of California, 1979), p. 87.

Appendices

APPENDIX A

MEASURING EARTHQUAKE MAGNITUDE AND INTENSITY

An earthquake's severity is generally expressed in two ways: (1) its size or magnitude, and (2) the intensity of damage and other effects.

Magnitude

The size of an earthquake and the amount of energy released is most commonly expressed using the *Richter magnitude scale.* An earthquake's Richter magnitude is determined by calculations based on the maximum amplitude registered on a standard instrument called a Wood-Anderson torsion seismograph.

The Richter magnitude varies logarithmically with an earthquake's maximum wave amplitude as seismographically recorded. Each whole number step upward on the magnitude scale represents an increase of ten times in the measured wave amplitude of an earthquake, and also represents approximately a 30-fold increase in energy released.

A quake that measures 2 on the Richter scale is the smallest normally felt by humans. Those with a Richter magnitude of 7 or more are commonly considered to be major earthquakes. The 1971 San Fernando, California, earthquake had a magnitude of 6.5. The 1906 San Francisco earthquake had a Richter magnitude of 8.3, and the 1964 Alaskan earthquake was rated 8.6. The largest earthquakes for which we have records have been about 8.9 on the Richter scale, which has no fixed maximum.

Intensity

The intensities of damage and other earthquake effects are expressed by the *Modified Mercalli intensity scale*, a measure that is partly subjective, depending on observations of what happens during an earthquake. Magnitude and intensity are not directly comparable: each earthquake's magnitude is represented by a *single* Richter number, whereas one earthquake can produce many different intensity measurements, depending on ground and structural conditions in various affected areas, and also on distances from the earthquake's epicenter.

Despite their lack of direct comparability, however, there are some rough relationships between an earthquake's Richter magnitude, the Modified Mercalli intensities to be expected near its source, and the peak ground accelerations likely to be experienced. Bruce Bolt notes that "magnitudes of earthquakes are used in an approximate way to predict what the greatest acceleration of the ground shaking may be in an earthquake site of an important structure," and that engineers use the information in designing structures to withstand such strong motions. Bolt adds a caveat: "This practice is somewhat unfortunate, however, because near an earthquake there is no strong correlation between earthquake magnitude and maximum peaks of acceleration."

In any event, the following tabulation shows approximate relationships between an earthquake's Richter magnitude and the Modified Mercalli intensities *to be expected near the source*, and Appendix B gives approximate relationships between Modified Mercalli intensity numbers and the average peak accelerations likely to be associated with such damage.

Magnitude and Intensity: Approximate Relationships

Magnitude	Intensity: Maximum Effects Expected Near the Source	
Richter Number[a]	Modified Mercalli Number[a]	Description of Effects Near the Source[b]
2	I-II	Usually detected only by instruments
3	III	Felt indoors
4	IV-V	Felt by most people; slight damage
5	VI-VII	Felt by all; many frightened and run outdoors; damage minor to moderate
6	VII-VIII	Damage moderate to major
7	IX-X	Major damage
8+	X-XII	Major damage and total damage

[a]*Note*: By convention and to avoid confusion, the Richter magnitude scale uses Arabic numbers, while the Modified Mercalli intensity scale is expressed in Roman numerals.

[b]See Appendix B for a more detailed statement of the Modified Mercalli intensity scale.

Sources: This description of earthquake measurements is adapted from Charles F. Richter, *Elementary Seismology* (1958); Bruce A. Bolt, *Earthquakes: A Primer* (1978), pp. 102-107; and California Division of Mines and Geology, "How Earthquakes are Measured," *California Geology* 32 (2): 35-37 (February 1979), pp. 35-37.

APPENDIX B

MODIFIED MERCALLI INTENSITY SCALE

In 1902 an Italian seismologist named Mercalli devised an earthquake intensity scale with a I to XII range. Mercalli's scale was modified in 1931 by American seismologists Wood and Neumann, to take modern structural features into account. It has since been reworked further, and several slightly different but basically similar versions are now found in the literature. The scale and descriptions given below draw upon a number of these sources.

The Modified Mercalli intensity scale measures the intensity of an earthquake's effects in a given locality, and is perhaps especially meaningful to the layman because it is based on actual observations of earthquake effects at specific places.

Modified Mercalli Intensity Number	Descriptive Term	Observed Effects	Average Peak Acceleration (g is gravity)
I	Imperceptible	Not felt at all, or only by a few under especially favorable circumstances.	
II	Very slight	Felt indoors but only on upper floors by persons at rest or favorably placed. Delicately suspended objects may swing.	
III	Slight	Felt noticeably indoors. Vibrations like those caused by light trucks. May not be recognized as an earthquake. Hanging objects swing.	
IV	Moderate	Felt by many indoors, but by few outdoors. Some sleepers awakened. Hanging objects swing. Vibration like that caused by heavy trucks, or a jolt as if a heavy object had struck the building. Parked cars rocked noticeably. Windows, doors, dishes and crockery rattle. Glasses clink. In the upper range of IV, wooden walls and frames creak.	0.015-0.02g

(continued)

Modified Mercalli Intensity Number	Descriptive Term	Observed Effects	Average Peak Acceleration (g is gravity)
V	Fairly strong	Felt by nearly everyone. Many sleeping persons awakened. Small unanchored objects displaced or overturned. Liquids disturbed, some spilled. Shutters and pictures set in motion. Doors open and close. Pendulum clocks may stop.	0.03-0.04g
VI	Strong	Felt by all. Many frightened and some may run outdoors. Walking is difficult and people walk unsteadily. Windows, dishes, crockery and glassware break. Books and knick-knacks fall from shelves. Pictures fall from walls. Furniture moves or overturns. Cracks may appear in weak plaster and materials of construction quality D. Damage slight. Small church or school bells ring. Trees and bushes visibly shaken or heard to rustle.	0.06-0.07g
VII	Very strong	Difficult for people to stand. Noticed by car drivers and passengers. Hanging objects quiver. Furniture is broken. Weak chimneys break at roof level. Fall of plaster, loose bricks, stones, tiles, cornices, unbraced parapets and architectural ornaments. Materials of construction quality D sustain serious damage. In some cases, cracks appear in materials of construction quality C. Waves appear on ponds, and water is made turbid. Small slides and cave-ins along sand or gravel banks. Large bells ring. Concrete irrigation ditches damaged.	0.10-0.15g

(continued)

Modified Mercalli Intensity Number	Descriptive Term	Observed Effects	Average Peak Acceleration (g is gravity)
VIII	Destructive	Persons driving cars disturbed. Fall of stucco and some masonry walls. Chimneys, monuments, towers and raised tanks collapse. Frame houses moved on foundations if not securely anchored. Panel walls thrown out of frame structures. Very heavy damage to materials of construction quality D, and some damage to materials of construction quality C, including partial collapse. Some damage to material of construction quality B. Branches broken from trees. Changes in flow or temperature of springs. Changes in water level of wells. Cracks appear in moist ground and on steep slopes.	0.25-0.30g
IX	Highly destructive	General panic. Unanchored frame structures shifted off foundations, or collapse. Well-designed frame structures thrown out of plumb. Loadbearing members of reinforced concrete structures cracked. Damage great in substantial buildings, with partial collapse. General damage to foundations. Materials of construction quality D completely destroyed. Serious damage to materials of construction quality C, and frequent collapse. Serious damage also sustained by material of construction quality B. Underground pipes broken. Conspicuous cracks appear in the ground. Serious damage to reservoirs. Water, sand and mud ejected in alluvial areas.	0.50-0.55g

(continued)

Modified Mercalli Intensity Number	Descriptive Term	Observed Effects	Average Peak Acceleration (g is gravity)
X	Extremely destructive	Most masonry and frame structures destroyed, along with their foundations. Some well-built wooden structures destroyed. Reinforced steel buildings and bridges seriously damaged, and some destroyed. Serious or severe damage to dams, dikes, weirs, and embankments. Substantial landslides. Sand and mud shifted horizontally on beaches and flat land. Rails bent.	More than 0.60g
XI	Disaster	Virtually all structures collapse. Only a few buildings left standing. Large, well-constructed bridges destroyed or severely damaged. Rails greatly bent and thrown out of position. Earth slumps and land slips in soft ground. Underground pipelines completely out of service.	More than 0.60g
XII	Major disaster	Damage nearly total. Practically all works of construction greatly damaged or destroyed. Objects thrown into the air. Waves seen on ground surface. Lines of sight distorted. Surface and underground streams changed in many ways. River courses altered.	More than 0.60g

Key: see page 76.

Sources: Adapted from Charles F. Richter, *Elementary Seismology* (1958), pp. 136-138; Bruce A. Bolt, *Earthquakes: A Primer* (1978), pp. 102-107; California, Division of Mines and Geology, "How Earthquakes Are Measured," *California Geology* 32 (2): 35-37 (February 1979); and Reinsurance Offices Association, *Earthquake Study: Mexico* (November 1977), p. 21.

Key: Construction quality A, B, C, and D, referred to in the intensity descriptions, are listed below in order of the degree of earthquake resistance provided. For example, construction quality A provides the greatest earthquake resistance, whereas construction quality D provides very poor resistance.

A. Construction quality A includes good workmanship, mortar and design; reinforcement, especially lateral, bound together using steel, concrete, etc.; designed to resist lateral forces (sideways shaking).

B. Construction quality B includes good workmanship and mortar; has reinforcement, but not designed to resist strong lateral forces.

C. Construction quality C includes ordinary workmanship and mortar; no extreme weaknesses, such as failing to tie in at corners, but not designed or reinforced to resist lateral forces.

D. Construction quality D includes low standards of workmanship; weak materials such as adobe and poor mortar; structures are weak in resistance to lateral forces.

APPENDIX C

RECOMMENDED EMERGENCY AND DISASTER ACTIVITIES

This is a detailed listing of emergency preparedness measures that can appropriately be taken in a state such as California that experiences significant seismic activity. The listing is based on the 1974 report of the Advisory Group on Disaster Preparedness to the California Legislature's Joint Committee on Seismic Safety.*

Disaster Planning

1. A compendium of legislation of the California Emergency Plan should be available to all potential users, with changes and corrections made annually.
2. The state should insure that current and comprehensive local emergency ordinances are in existence in all local jurisdictions.
3. The intergovernmental mutual aid programs should be strengthened and their applicability broadened.
4. Barriers to effective use of resources in disasters should be identified, and plans made to remove them. Such barriers include jurisdictional, legal and procedural hurdles.
5. The State Office of Emergency Services should be empowered to give incentives to local governments and organizations that prepare and maintain comprehensive emergency preparedness plans.
6. Counties should be designated as coordinators of intracounty emergency operations, unless there are better alternatives in specific areas.
7. A sound and thorough national disaster preparedness program should be established.
8. Emergency action checklists and standard operating procedures should be developed by all agencies that perform emergency safety functions.
9. A statewide emergency medical care plan should be developed, with (a) a public health section, and (b) a patient care section.
10. Both private and public schools should be required to have disaster plans, and better means should be developed to evaluate their quality.
11. Disaster plans should include procedures for identification and control of "official" visitors to earthquake damaged areas. The entry of persons whose presence is not warranted could thus be discouraged, and the work of those who need to have access facilitated.

---

*California, Legislature, Joint Committee on Seismic Safety, *Meeting the Earthquake Challenge* (1974), pp. 57-96.

12. Designs for all buildings over eight stories high should be required to have an acceptable plan for evacuating occupants, or for creating safe internal environments in case of earthquake or fire.

13. Disaster plans should include means for controlling the solicitation and donation of articles not needed for disaster relief, as well as procedures for efficient handling of the articles received.

14. Organizations having emergency service functions should be required to have a plan for the effective use of available communication facilities in an earthquake disaster.

Strengthening Operating Capabilities

1. State and local governments should develop and enforce standards for earthquake-resistant construction of emergency service structures and associated non-structural elements.

2. State law should require maintenance and safe storage of duplicate, back-up records that may be needed to repair and restore vital utility services.

3. When stockpiled, emergency equipment and supplies should not be stored in hazardous structures.

4. Supplies of "Unsafe Building" signs should be prepared and stored for use after an earthquake.

5. The State Emergency Plan should include a workable system for gathering thorough and consistent information and statistical data about disasters and their effects.

6. The state should have adequate programs to train local safety and public works employees in emergency rescue operations.

7. Regular fire and earthquake drills should be required in all elementary, intermediate and secondary schools.

8. An earthquake education program should be developed to (a) inform the public about earthquakes, feasible precautions, and appropriate behavior, and (b) further public understanding of the need for state seismic safety programs.

9. Appropriations for emergency and disaster assistance should be greatly increased, and the funds made available to local jurisdictions for disaster assistance purposes. The performance-monitoring and emergency response overview roles of the State Office of Emergency Services should be strengthened.

10. All services available in connection with disasters should be summarized in widely disseminated informational charts, which should be regularly reviewed and updated.

11. Federal disaster assistance programs should be reviewed and where necessary modified to insure equity, consistency, and effective operation.

12. Local governments with the capability of doing so should be encouraged to develop special staffs to be responsible for planning for all types of major contingencies, including disasters. Many local governments are likely to need state and perhaps federal assistance if adequate disaster planning is to be done.

Evaluating Performance

1. Programs should be initiated to develop criteria for evaluating the adequacy of disaster preparedness efforts, prior to an actual disaster.
2. There should be a formal process for review and approval of local emergency plans by the State Office of Emergency Services.
3. Annual disaster exercises should be conducted, involving state and federal agencies, local governments, and the larger private organizations.
4. The Office of Emergency Services should be responsible for systematic analysis and on-site evaluation of disaster operations and recovery activities during and immediately after major disasters.

Minimizing Future Disaster Problems

1. Building and safety codes should be improved in the interest of earthquake safety.
2. The application and enforcement of codes and other earthquake laws and ordinances should be improved. This should be a responsibility of the Seismic Safety Commission or other appropriate state agency.
3. The state should require local jurisdictions to establish programs for the orderly abatement of earthquake hazards, especially in buildings constructed before adoption of California's Riley Act in 1933.
4. State requirements should insure that small privately owned utility systems, such as those in planned unit developments, are properly designed, managed, and maintained.
5. Minimum standards for various kinds of public safety services should help insure that local governments and other jurisdictions have resources adequate to meet anticipated emergency needs.
6. State standards should provide for a periodic review by local governments of the occupancy of structures. This should be closely coordinated with active hazardous-building abatement programs.
7. Field Act provisions should be extended to privately owned school buildings, and state college and university buildings.

APPENDIX D

AUBURN DAM SEISMIC SAFETY REVIEW

(Text of Policy Statement Adopted March 10, 1977 by the California Seismic Safety Commission)

The Seismic Safety Commission has conducted a field inspection of the Auburn Dam site and reviewed site preparations completed to date. It has considered written and oral testimony with regard to the seismic design of the dam and its safety during earthquake shaking. It has heard a progress report on the work of consultants employed by the Bureau of Reclamation to review the seismic hazards of the site. The progress report did not, however, relate to the safety of the dam design.

On the basis of this and other information, we conclude that a significant seismic safety question exists, whose resolution will require careful consideration after June 1, 1977, of the consultant's report on Auburn Dam promised on or about that date and of the findings of the special panel of experts appointed by the Bureau to review and comment on the geologic and seismic hazards to which the proposed dam will be exposed.

Accordingly, we recommend that the State of California conduct an independent review of the proposed dam's seismic safety, employing sufficient funds and staff to pursue a thorough investigation and complete it expeditiously. We also recommend that the Bureau of Reclamation and Department of Interior avoid any conflicting commitments respecting the dam's future until the independent review by the State of California has been concluded and its findings made available to the public.

To that end, we recommend that the Governor of California designate a lead agency to have principal responsibility for the safety review, which should be conducted in accordance with the procedures and policies specified herein. We also recommend that the Governor and other agencies and representatives of the citizens of California explore with appropriate federal authorities the early implementation of the safety review policies and procedures outlined below:

1. A Lead Agency

The Governor should designate a State lead agency in order to focus responsibility for the dam safety investigation, as well as for marshalling the full range of resources necessary to insure a careful and thorough review, begun promptly and conducted with reasonable dispatch. We believe that the lead agency for the Auburn Dam review and any other dam reviews should be the Division of Safety of Dams, Department of Water Resources and that the responsible person should be the Director of that department.

In conducting its review, the lead agency should call upon the services and assistance of other appropriate State agencies, and especially upon the Office of the State Geologist. The State Geologist is responsible for the collection and dissemination of data relating to geologic hazards and seismic safety. Consequently, it is imperative that the lead agency coordinate its efforts with the Office of the State Geologist, and utilize the data collected and submitted by that office.

2. A Thorough and Expeditious Review

The State lead agency should conduct a thorough and expeditious review of the design of the projected dam and of all relevant geologic, seismic, engineering, and other data in order to reach a conclusion as to whether or not the future safety of the citizens of California is adequately protected against failures caused by earthquake shaking or other seismic activity.

3. Using the Resources California Has Available

The magnitude of Auburn Dam will require the lead agency to call on the engineering and scientific resources available in the universities and private sector in addition to those of the State and federal governments. Fortunately, California has excellent human and technological resources for the study of earthquake related phenomena, as well as other aspects of dam safety. Appropriate use must be made of these resources in conducting a first-rate review of the dam's safety. Consequently, these resources should be employed to complement the currently available dam review resources of the State government.

Because of our special concern with the seismic safety of dams constructed in California, we recommend that a State review board be appointed to examine and report on the safety of Auburn Dam. The board should comprise individuals who are highly qualified to evaluate all aspects of dam safety, including seismic hazard, and include at least one geoscientist, and at least one engineer having recognized qualifications and experience that insure familiarity with the latest design procedures and requirements for dams. It is proposed that such members appointed to review the seismic aspects of the hazard be selected with the advice and approval of the Seismic Safety Commission.

4. Time and Money

To recapitulate, the review should be both thorough and expeditious, and ample funds should be provided for this purpose. We believe that a complete and independent analysis should be performed, similar to the independent State review performed by the Office of the State Architect with respect to the safety of hospitals and schools. We believe that a thorough review of a project as large and important as Auburn Dam will require the commitment of resources of the order of \$200,000 to \$500,000. Such expenditures for design review are not unusual for special studies of this nature, and even the top figure quoted would approximate less than one-tenth of one percent of the estimated cost of the dam. It may be possible to complete a well-financed review within six months, and in any event the work should be finished in not more than one year assuming good cooperative arrangements with the appropriate federal agencies concerned with the dam's design. The State's safety review of Auburn Dam should begin *immediately*.

5. Progress Reports

During the review, the lead agency should make periodic progress reports at least bimonthly, and more frequently if appropriate, to the Governor and the Seismic Safety Commission.

6. Avoiding Irreversible Commitments During the Review

The Bureau of Reclamation and the Department of Interior should be asked to scrupulously avoid making any irreversible commitments respecting the future construction of the dam until the State's independent review has been completed. Otherwise the purpose of the review may be defeated before its final results are known.

7. Independent Review of Other Federal Dam and Reservoir Projects

Our studies also lead to the conclusion that all future federal dam and reservoir projects in California should receive a similar independent, thorough, and expeditious review by the State, conducted with the aid of independent panels of highly qualified professionals who are answerable to the State's lead agency and to the Seismic Safety Commission.

Independent State review of federal projects appears highly appropriate for the following reasons:

First, the principle of independent review and checking of all designs of projects that could create future hazards is a well-established principle of good practice, and a necessary measure for protecting the public safety. Accordingly, the principle of independent review must be incorporated into State and federal policy.

Second, the Teton Dam disaster and other data make it clear that enterprises on the scale of most federal water projects do have the potential of creating major hazards, and illustrate the danger of omitting a complete and independent review of site hazards and design safety.

Third, the State government has a fundamental responsibility for taking all reasonable measures to guarantee the safety of its citizens. Moreover, the State of California–speaking for the people living in one of the nation's most seismically active geographic regions–is the most appropriate agency of representative government to review projects having the potential of creating major hazards for its citizenry and economy, especially during earthquakes.

Fourth, California is fortunate to have probably the best human and technological resources available anywhere in the world to deal with questions of seismic safety. Accordingly, it appears appropriate that the State muster those resources in reviewing the seismic safety of potentially hazardous projects.

Fifth, the principle of State review of federal plans, and of federal agency compliance or consistency with State objectives, is an important policy that underlies much recent thinking on improved federal-state relationships in the public interest.

Sixth, a competent and professional engineering organization should welcome an independent and unbiased review of data, assumptions and procedures, especially with respect to unique, monumental structures that involve substantial public safety considerations. Such practices can help identify and correct design errors, or alternatively can help confirm that designs insure adequate margins of safety. In short, the practice of independent review brings to bear the judgments of additional highly qualified experts which helps to guarantee achievement of acceptable safety levels.

8. Independent Review of Non-Federal Dams in California

In addition to independent State review of federally designed dams, there is also a need for a special independent review of the safety, including seismic safety, of dams in California, including those designed by the State Department of Water Resources. Under current practice, there are independent

State reviews of privately designed dams. Similarly, dams designed by State agencies should also receive a complete and independent design safety review.*

We believe that all the reviews of significant dams in California should be conducted by special boards established to assure the adequacy of all aspects of the designs, including their seismic safety. As in the case of Auburn Dam, our special concern with seismic safety prompts us to recommend that all such review boards comprise individuals who are highly qualified to evaluate all aspects of dam safety, including seismic and geologic hazard, and include at least one geoscientist, and at least one engineer having recognized qualifications and experience that insures familiarity with the latest seismic design procedures and requirements for dams. It is proposed that such members appointed to review the seismic aspects of the hazard be selected with the advice and approval of the Seismic Safety Commission.

CONCLUSION

We believe that adoption and implementation of the safety policies and review measures outlined above is an essential safeguard relating to the future security of the residents of California. Accordingly, we urge each of the officials and agencies who have been named to pursue appropriate actions starting immediately. We especially urge the Governor to consider the imperative need for early and expeditious action in conducting a complete and independent review of the safety of Auburn Dam.

---

*Author's Note: Since passage of this resolution, it has been pointed out that the Division of Safety of Dams does review the safety of all dams designed by state agencies. It is true that the California Water Code, Sec. 6005 explicitly includes dams owned by all state departments, agencies and subdivisions under the review and surveillance umbrella of the division. On the other hand, the Division of Safety of Dams is a subsidiary agency within the Department of Water Resources, which in turn is the principal designer, owner and operator of state-owned dams in California. Accordingly the author believes the Seismic Safety Commission had this close intra-agency relationship in mind when arguing for a "complete and independent" review of the safety of dams designed by state agencies. In short, there may have been some concern that a close relationship between the design agency and the review agency may not fully comply with the principle of *independent* design review. On the other hand, it should be acknowledged that the design of dams owned by the Department of Water Resources must be reviewed by a board of three consultants retained by the department to make independent findings of the safety of dams proposed to be built or enlarged by the department. (*California Water Code*, Sec. 6056, and *California Administrative Code*, Secs. 340-343.)

Bibliography

Allen, Clarence R.
"Geological Criteria for Evaluating Seismicity." C. Lomnitz and E. Rosenblueth, eds. *Seismic Risk and Engineering Decisions.* Amsterdam: Elsevier Scientific Publishing Co., 1976, pp. 31-69.

American Institute of Architects Research Corporation.
Architects and Earthquakes: Research Needs. Washington, D.C.: n.d., 247pp.

Summer Seismic Institute for Architectural Faculty. August 1-12, 1977. Washington, D.C.: October 1977. 321pp.

American Society of Civil Engineers.
The Current State of Knowledge of Lifeline Earthquake Engineering. ASCE Specialty Conference. Proceedings. Los Angeles: University of California, August 30, 1977. 478pp.

Association of Bay Area Governments.
Earthquake Insurance Issues. Berkeley: September 1977. n.p.

Earthquake Intensity and Expected Cost in the San Francisco Bay Area. [title of file] Berkeley: February 1978. File of five maps and accompanying text. 12pp.

Hazards Evaluation for Disaster Preparedness Planning. Berkeley: February 1976. 37pp.

Land Capability Analysis for Planning and Decision Making. Berkeley: February 1976. 30pp.

Legal References on Earthquake Hazards and Local Government Liability. Berkeley: December 1978. 189pp.

Regional Earthquake Safety Issues and Objectives. Berkeley: 1977. 11pp.

San Francisco Bay Region: A Review of Local Regulations Related to Geologic and Hydrologic Hazards, Constraints and Resources. Berkeley: May 1977. 8pp.

Blume, John A.

"Civil Structures and Public Safety: Safety in Design and Construction of Civil Structures." *Public Safety: A Growing Factor in Modern Design*. Symposium. National Academy of Engineering. Washington, D.C.: 1970. 115pp.

"Civil Structures and Earthquake Safety." *Earthquake Risk*. Conference Proceedings, September 22-24, 1971. California. Legislature. Joint Committee on Seismic Safety. Sacramento: 1972. 27pp

Bolt, Bruce A.

Earthquakes: A Primer. San Francisco: W.H. Freeman and Company, 1978. 241pp.

"Earthquake Hazards." *EOS: Transactions, American Geophysical Union*, 59(11) 946-962 (November 1978).

California. Department of Conservation. Division of Mines and Geology.

Compilation of Strong-Motion Records Recovered from the Santa Barbara Earthquake of 13 August, 1978. L.D. Porter. Preliminary Report 22. Sacramento: October 20, 1978. 43pp.

Fault Hazard Zones in California. Special Publication 42. Earl W. Hart. Revised January 1977. 24pp.

CDMG Note Number 37, *Guidelines to Geologic/Seismic Reports*, 1p. CDMG Note Number 43, *Recommended Guidelines for Determining the Maximum Credible and the Maximum Probable Earthquakes*, 1p. CDMG Note Number 44, *Recommended Guidelines for Preparing Engineering Geologic Reports*, 1p. CDMG Note Number 46, *Guidelines for Geologic/Seismic Considerations in Environmental Impact Reports*, 1p. CDMG Note Number 49, *Guidelines for Evaluating the Hazard of Surface Fault Rupture*, 2pp. CDMG Note Number 53, *California Well Sample Respository Operating Procedures and Policies*, 1p. Sacramento: August 1973-September 1976.

Urban Geology: Master Plan for California. Bulletin 198. Sacramento: 1973. 112pp.

California. Department of Conservation. Governor's Earthquake Council.

Second Report of the Governor's Earthquake Council. Sacramento: September 1974. 86pp. (This report is available from the California Division of Mines and Geology as Special Publication 46.)

California. Department of General Services. Office of Architecture and Construction. Fred W. Cheesebrough.

Letter with enclosures containing a brief description of the Field Act, the Garrison Act, the Riley Act, Title 21, Title 24 and the Uniform Building Code, and forms for filing Administrative Regulations with the Secretary of State. February 3, 1976. v.p.

California. Department of General Services. Office of the State Architect.

Notice of Proposed Changes in the Regulations of the Department of General Services, Office of the State Architect. Fred W. Cheesebrough. Sacramento: August 22, 1977. 29pp.

[California. Department of Transportation.] Earthquake Committee.
"Commentary on Earthquake Design Criteria." (January 1978, Memo 15-10) supplements to the *Earthquake Design Criteria, Bridge Planning and Design Manual*, v. 1, secs. 2-16 (April 1977).

California. Department of Water Resources.
"California's Dam Surveillance Program." Gordon W. Dukleth. Paper presented at Engineering Foundation Conference, "The Evaluation of Dam Safety." Asilomar Conference Grounds, Pacific Gove, California: November 30, 1976. 10pp.

"Effective State Dam Safety Supervision." Gordon W. Dukleth. Presented at Seminar on New Perspectives on the Safety of Dams, Stanford University, August 28, 1978. 9pp.

"Report on States' Experiences and Their Problems in the Inspection, Maintenance and Rehabilitation of Old Dams." Clifford J. Cortright. Presented to the Engineering Foundation Conference, September 23-28, 1973, Pacific Grove, California. 13pp.

"The Dam Safety Program." Gordon W. Dukleth. Presented to the California Water Commission in Stockton, California, February 4, 1977. 11pp.

Investigation of Failure: Baldwin Hills Reservoir. Sacramento: April 1964. 64pp. + photographs and plates.

California. Department of Water Resources. Division of Safety of Dams.
Statutes and Regulations Pertaining to Supervision of Dams and Reservoirs 1978. Sacramento: 1978 [pamphlet] 45pp.

California. Legislature. Joint Committee on Seismic Safety.
Final Report to the Legislature. *Meeting the Earthquake Challenge*. Sacramento: January 1974. 223pp. (This report is available from the California Division of Mines and Geology as Special Publication 45.)

California. Legislature. Joint Committee on Seismic Safety. Special Subcommittee.
The San Fernando Earthquake of February 9, 1971 and Public Policy. George O. Gates, ed. Sacramento: July 1972. 127pp.

California. Office of Emergency Services.
Earthquake Response Plan. Sacramento: November 1, 1977. v.p.

Report to the Seismic Safety Commission. [On dam safety and inundation map program] Everett F. Blizzard. Sacramento: June 10, 1976. 20pp.

California. Office of Emergency Services. California Earthquake Prediction Evaluation Council.
Earthquake Prediction Evaluation Guidelines. Sacramento: February 1977. 9pp.

[California. Office of Emergency Services. California Earthquake Prediction Evaluation Council.]
"Earthquake Prediction Evaluation Guidelines," *California Geology*, 30:158-160 (July 1977).

[California.] Resources Agency. Geologic Hazards Advisory Committees for Program and Organization.

Earthquake and Geologic Hazards in California: A Report to the Resources Agency. Sacramento: April 26, 1967. 18pp.

California. Seismic Safety Commission.

A Review of the Seismic Safety Element Requirement in California. Prepared by the Seismic Safety Element Review Committee. SSC 77-01. Sacramento: March 25, 1977. 23pp. + appendices.

The Field Act and California Schools. SSC 79-02. Sacramento: March 1979. 43pp. + appendices.

Field Act Review. Sacramento: December 12, 1976. Draft. 57pp. + appendix.

Goals and Policies for Earthquake Safety in California. Stanley Scott and Robert Olson . . . . SSC 79-04. Sacramento: May 10, 1979. 64pp. + references.

Hazardous Buildings: Local Programs to Improve Life Safety. SSC 79-03. Sacramento: March 8, 1979. 30pp. + appendices.

Legislative Digest. Sacramento: May 10, 1976. 18pp.

Memorandum. *Workshop Summary Report.* Seismic Safety Commission Workshop on New Building Construction, Critical Facilities and Energy and Lifeline Systems, and Community Planning and Development, September 1976. Sacramento: March 10, 1977. v.p.

Memorandum. *Workshop Summary Reports.* Robert A. Olson, Executive Director. Sacramento: March 10, 1977. 20pp.

Report on Local Government Implementation of the Special Studies Zones Act. Seismic Safety Commission Staff. Sacramento: July 1977. 50pp.

Report of the Task Committee of the Seismic Safety Commission on the Hospital Act of 1972. SSC 77-03. Sacramento: May 12, 1977. 22pp. + appendices.

Report on State Agency Programs for Seismic Safety. Christopher J. Wiley. Sacramento: April 30, 1979. Draft. 163pp.

Cluff, Lloyd S.

"Geologic Perspectives on Earthquake Hazards and Dam Safety." Woodward-Clyde Consultants, lecture notes for Seminar on New Perspectives on Safety of Dams. Stanford University, August 28-September 1, 1978. 15pp.

Cortright, Clifford J.

A New Look at California Dams and Earthquakes. American Society of Civil Engineers, Meeting Preprint 2011. San Francisco: April 9-13, 1973. 10pp.

"Reevaluation and Reconstruction of California Dams." *Journal of the Power Division*, 55-72. ASCE, 96, no. P01, Proc. Paper 7008 (January 1970).

Degenkolb, Oris H.

Bridge Earthquake Restrainer Program. Sacramento: Department of Transportation, April 1978. 5pp.

Retrofitting Techniques for Highway Bridges. Sacramento: Department of Transportation, n.d. 13pp.

Summary of Highway Damage–August 13, 1978. [Sacramento: Department of Transportation, n.d.] 4pp.

Donovan, Neville C., Bruce A. Bolt, and Robert V. Whitman.

"Development of Expectancy Maps and Risk Analysis." *Journal of the Structural Division*, 1179-1192. Proceedings of the American Society of Civil Engineers, 104, no. ST8, Proc. Paper 13972 (August 1978).

Dukleth, Gordon W. and E.W. Stroppini.

"Seismic Safety for California Dams." *California Geology*, 29:243-248 (November 1976) (a publication of the Division of Mines and Geology).

Earthquake Engineering Research Institute.

Earthquake Response Procedures. Berkeley: EERI, revised October 1978. v.p.

Learning from Earthquakes: 1977 Planning and Field Guides. Berkeley: EERI, 1977. 200pp.

Evans, John.

Attorney's Guide to Earthquake Liability. Berkeley: Association of Bay Area Governments, 1979 (in press).

Haas, J. Eugene and Dennis S. Mileti.

Socioeconomic Impact of Earthquake Prediction on Government, Business, and Community: Research Findings, Issues and Implications for Organizational Policy. Boulder: University of Colorado, Institute of Behavioral Science, n.d. 30pp.

Howell, B.F., Jr.

"Average Regional Seismic Hazard Index (ARSHI) in the United States." Douglas E. Morgan et al., eds. *Geology, Seismicity, and Environmental Impact.* Association of Engineering Geologists. Los Angeles: University Publishers, Special Publications, October 1973. pp. 277-285

Howell, B.F., Jr. and T.R. Schultz.

"Attenuation of Modified Mercalli Intensity with Distance from the Epicenter." *Bulletin of the Seismological Society of America*, 65(3)651-665 (June 1975).

Kerr, Richard A.

"U.S. Earthquake Hazards: Real but Uncertain in the East." *Science*, 201:1001-1003 (September 15, 1978).

Los Angeles. City.
Consensus Report of the Task Force on Earthquake Prediction. Rachel Gulliver Dunne, Chairman. Presented to Mayor Tom Bradley. Los Angeles: October 1978. 47pp.

[Mancarti, Guy D.]
Maintaining Transportation Lifelines. Sacramento: [California Department of Transportation] [July 1977]. 9pp.

Mancarti, Guy D. and Oris H. Degenkolb.
Retrofitting Structures to Increase Seismic Resistance of Highway Lifelines. Sacramento: California Department of Transportation, n.d. 15pp.

Margerum, Terry.
"Earthquake Hazards and Local Government Liability: Executive Summary." Berkeley: Association of Bay Area Governments. n.d., n.p.

Will Local Government be Liable for Earthquake Losses? What Cities and Counties Should Know About Earthquake Hazards and Local Government Liability. Berkeley: Association of Bay Area Governments, January 1979. 32pp.

Meltsner, A.J.
Seismic Safety of Existing Buildings and Incentives for Hazard Mitigation in San Francisco: An Exploratory Study. Berkeley: Earthquake Engineering Research Center, University of California, December 1977. 79pp.

National Research Council. Assembly of Mathematical and Physical Sciences. Committee on Seismology. Panel on Earthquake Prediction.
Predicting Earthquakes: A Scientific and Technical Evaluation–with Implications for Society. Washington, D.C.: National Academy of Sciences, 1976. 62pp.

National Research Council. Commission on Sociotechnical Systems. Advisory Committee on Emergency Planning. Panel on the Public Policy Implications of Earthquake Prediction.
Earthquake Prediction and Public Policy. Washington, D.C.: National Academy of Sciences, 1975. 142pp.

National Research Council. Committee on Socioeconomic Effects of Earthquake Predictions.
A Program of Studies on the Socioeconomic Effects of Earthquake Predictions. Washington, D.C.: National Academy of Sciences, 1978. 162pp.

National Science Foundation.
RANN-Research Applications Directorate and U.S. Department of the Interior. United States Geological Survey. *Earthquake Prediction and Hazard Mitigation Options for USGS and NSF Programs*. Washington, D.C.: September 15, 1976. 76pp.

Page, Robert A., John A. Blume, and William B. Joyner.
"Earthquake Shaking and Damage to Buildings." *Science*, 189(4203):601-608 (August 22, 1975).

Pate, Marie-Elisabeth.
Public Policy in Earthquake Effects Mitigation: Earthquake Engineering and Earthquake Prediction. Technical Report No. 30. Stanford: John A. Blume Earthquake Engineering Center, Stanford University, May 1978. 378pp.

Pate, Marie-Elisabeth and Haresh C. Shah.
"Public Policy Issues: Earthquake Prediction." Paper submitted for publication (presently available from the John A. Blume Earthquake Engineering Center, Stanford University, Stanford, Ca 94305).

Perkins, Jeannie.
Earthquake Intensity and Related Costs in the San Francisco Bay Area. Berkeley: Association of Bay Area Governments. February 1978. 12pp. (Explanatory text to accompany a file of five earthquake intensity and damage maps; see also reference above, under Association of Bay Area Governments).

Reed, Richard E., ed.
Living With Seismic Risk: Strategies for Urban Conservation: Proceedings of a Seminar. December 10, 1976, Casa Del Prado, Balboa Park, San Diego, California. Washington, D.C.: American Association for the Advancement of Science, 1977. 143pp.

Scott, Stanley.
"Earthquake and Geologic Safety Policies and Objectives." Unpublished paper. Berkeley: Institute of Governmental Studies, University of California, April 27, 1976. 20pp. + appendix.

Learning From San Fernando: Summary Report on the UCLA Earthquake Conference. Prepared for the Joint Committee on Seismic Safety, California State Legislature. Berkeley: Institute of Governmental Studies, University of California, August 1971. 20pp.

"Learning to Live With Earthquakes: Research and Policy for Seismic Safety in California." *Public Affairs Report*, 17(5):1-5. Berkeley: Institute of Governmental Studies, University of California, October 1976.

"Policies for Seismic Safety: Elements of an Intergovernmental Program." Unpublished paper. Berkeley: Institute of Governmental Studies, University of California, October 13, 1977. 42pp.

"Preparing for Earthquake Prediction." Unpublished paper. Berkeley: Institute of Governmental Studies, University of California, September 29, 1976. 10pp.

"Preparing for Future Earthquakes: Unfinished Business in the San Francisco Bay Area." *Public Affairs Report*, 9(6):1-5. Berkeley: Institute of Governmental Studies, University of California, December 1968.

"Towards a Partnership for Seismic Safety: 1974 and Beyond." Unpublished conference paper. Berkeley: Institute of Governmental Studies, University of California, April 26, 1972. 22pp. + appendices.

In the Interest of Earthquake Safety: Findings and Conclusions by Members of the Task Force on Earthquake Hazard Reduction, Office of Science and Technology, Executive Office of the President. Berkeley: Institute of Governmental Studies, University of California, 1971. 22pp.

Seed, H. Bolton.
"Dams and Soils." California. Joint Committee on Seismic Safety, Special Subcommittee. *San Fernando Earthquake of February 9, 1971 and Public Policy.* pp. 13-33. Sacramento: July 1972.

Shah, Haresh C., et al.
The Purpose and Effects of Earthquake Codes. Internal Study Report No. 1. Stanford: John A. Blume Earthquake Engineering Center, Stanford University, August 1977. 20pp + appendices.

Spangle, William, and Assoc.
Post-Earthquake Land Use Planning: Draft Final Report. Portola Valley, California: July 1979. 67pp + appendices.

Stanford Research Institute. Center for Resource and Environmental Systems Studies.
Earthquake Prediction in Society. Menlo Park, California: February 1977. 40pp.

Earthquake Prediction, Uncertainty, and Policies for the Future: A Technology Assessment of Earthquake Prediction. Prepared for the National Science Foundation. Menlo Park, California: January 1977. 315pp.

Steinbrugge, Karl V.
Earthquake Hazard in the San Francisco Bay Area: A Continuing Problem in Public Policy. Berkeley: Institute of Governmental Studies, University of California, 1968. 80pp.

Steinbrugge, Karl V. and Henry J. Degenkolb.
"Meeting the Earthquake Challenge: California's New Laws." *Civil Engineering* [ASCE] 45[2] :44-47 (February 1975).

Steinbrugge, Karl V. and Carl B. Johnson.
"Earthquake Hazard and Public Policy in California." *Engineering Issues*, 99(PP4): 513-519 (October 1973).

Structural Engineers Association of Northern California.
School Construction Under the Field Act. March 1953. 3pp.

Turner, Ralph H., Joanne M. Nigg, Denise Heller Paz, and Barbara Shaw Young.
Earthquake Threat: The Human Response in Southern California. Los Angeles: Institute for Social Science Research, University of California, 1979. 152pp.

U.S. Army. Chief of Engineers.
"Attenuation of High-Frequency Seismic Waves in the Central Mississippi Valley." Report 10, Miscellaneous Paper S-73-1. *State-of-the-Art for Assessing Earthquake Hazards in the United States.* Otto Nuttli and John J. Dwyer. Vicksburg, Miss.: U.S. Army Waterways Experiment Station, July 1978. 75pp.

"Specifying Peak Motions for Design Earthquakes." Report 7, Miscellaneous Paper S-73-1. *State-of-the-Art for Assessing Earthquake Hazards in the United States.* Ellis L. Krinitsky and Frank K. Chang. Vicksburg, Miss.: U.S. Army Waterways Experiment Station, December 1977. 9pp.

U.S. Committee on Large Dams of the International Commission on Large Dams.
Model Law for State Supervision of Safety of Dams and Reservoirs. New York: United States Committee on Large Dams [1970]. 29pp.

U.S. Department of Commerce. National Bureau of Standards.
Building Practices for Disaster Mitigation. Richard Wright, et al., eds. Building Science Series 46. Proceedings of a workshop . . . August 28-September 1, 1972. Washington, D.C.: February 1973. 474pp.

Research Needs and Priorities for Geotechnical Earthquake Engineering Applications. Kenneth L. Lee, et al., eds. A report of a workshop held at the University of Texas at Austin. Prepared for the National Science Foundation. Austin, Texas: June 1977. 133pp.

U.S. Department of Commerce. National Oceanic and Atmospheric Administration. National Geophysical and Solar-Terrestrial Data Center.
Final Report: An Analysis of Earthquake Intensities with Respect to Attenuation, Magnitude and Rate of Recurrence. Rutlage J. Brazee. NOAA Technical Memorandum EDS NGSCD-2. Boulder, Colorado: August 1976. Revised Edition. 53pp + appendices.

U.S. Department of the Interior. Geological Survey.
A Probabilistic Estimate of Maximum Acceleration in Rock in the Contiguous United States. S.T. Algermissen and David M. Perkins. Open File Report 76-416. Washington, D.C.: 1976. 45pp.

Earthquake Hazards Reduction Program-Fiscal Year 1978 Studies Supported by the U.S. Geological Survey. Robert M. Hamilton. Geological Survey Circular 780. Arlington, Virginia: 1978. 36pp.

Quantitative Land-Capability Analysis. Raymond T. Laird, et al. Association of Bay Area Governments. Geological Survey Professional Paper 945. Washington, D.C.: 1979. 115pp.

U.S. Executive Office of the President.
The National Earthquake Hazards Reduction Program. Washington, D.C.: June 22, 1978. 30pp.

U.S. Executive Office of the President. Office of Science and Technology. Task Force on Earthquake Hazard Reduction.
Earthquake Hazard Reduction: Program Priorities. Washington, D.C.: September 1970. 54pp.

U.S. Executive Office of the President. Office of Science and Technology. Working Group on Earthquake Hazards Reduction.
Earthquake Hazards Reduction: Issues for an Implementation Plan. Washington, D.C.: 1978. 231pp.

U.S. Library of Congress. Congressional Research Service.
Emergency Preparedness and Disaster Assistance: Federal Organization and Programs. Clark F. Norton. Washington, D.C.: April 18, 1978. 119pp.

Wooton, Tom M.
"Strong Motion Instrumentation Program." *California Geology*, 32 (4): 77-79 (April 1979).

RECENT INSTITUTE PUBLICATIONS

1979

Balderston, Frederick E., I. Michael Heyman and Wallace F. Smith
Proposition 13, Property Transfers, and the Real Estate Markets (prepared for the Commission on Government Reform). Research Report 79-1 56pp + Appendices $3.00

Bowen, Frank M. and Eugene C. Lee
Limiting State Spending: The Legislature or the Electorate (prepared for the Commission on Government Reform). Research Report 79-4 100pp + Appendices $4.50

Dean, Terry J. and Ronald J. Heckart, compilers
Proposition 13 in the 1978 California Primary: A Pre-Election Bibliography. Occasional Bibliographies Number 1. 88pp $6.00

Fletcher, Thomas, Dennis Hermanson, John Taylor, Shirley Hentzell and Dean Linebarger
Allocating the One Percent Local Property Tax in California: An Analysis (prepared for the Commission on Government Reform). Research Report 79-2 35pp + Appendices $3.25

MacWatters, Ann Robertson
Financing Capital Formation for Local Governments (prepared for the Commission on Government Reform). Research Report 79-3 51pp $3.00

1978

Ames, Bruce N.
Environmental Chemicals Causing Cancer and Genetic Birth Defects: Developing a Strategy to Minimize Human Exposure. California Policy Seminar Monograph Number 2. 28pp $3.50

Bradshaw, Ted K. and Edward J. Blakely
Policy Implications of California's Changing Life Styles. California Policy Seminar Monograph Number 3. 30pp $3.50

Cooper, Michael D.
California's Demand for Librarians: Projecting Future Requirements. 128pp $6.50

Eckbo, Garrett
Public Landscape: Six Essays on Government and Environmental Design in the San Francisco Bay Area. 135pp $7.75

Gates, Paul W., Ronald B. Robie, John T. Knox and Norman Y. Mineta
Four Persistent Issues: Essays on California's Land Ownership Concentration, Water Deficits, Sub-State Regionalism, and Congressional Leadership. 79pp $5.75

Mullins, Phil, Thomas O. Leatherwood and Arthur Lipow, eds.
Political Reform in California: Evaluation and Perspective. Research Report 78-3 160pp $6.50

Nathan, Harriet and Stanley Scott, eds., with 13 local authors
Experiment and Change in Berkeley: Essays on City Politics, 1950-1975. 501pp $10.75

Phillips, Kenneth F. and Michael B. Teitz
Housing Conservation in Older Urban Areas: A Mortgage Insurance Approach. Research Report 78-2 39pp $3.50

Pyle, David H.
Changes in the Financial Services Industry in California. California Policy Seminar Monograph Number 1. 17pp $3.00

Stebbins, Robert C., Theodore J. Papenfuss and Florence D. Amamoto
Teaching and Research in the California Desert. Research Report 78-1 26pp $3.00

Weeks, Kent M.
Ombudsmen Around the World: A Comparative Chart. 2d. ed. 163pp $7.50

1977

Burns, Eveline M.
Social Welfare in the 1980s and Beyond. 20pp $4.00

Capell, Elizabeth A.
Constitutional Officers, Agencies, Boards and Commissions in California State Government, 1849 to 1975. Research Report 77-1 61pp $3.50

Costonis, John J., Curtis J. Berger and Stanley Scott
Regulation v. Compensation in Land Use Control: A Recommended Accommodation, A Critique, and an Interpretation. 91pp $3.50

Eells, John M.
LAFCO Spheres of Influence: Effective Planning for the Urban Fringe. Working Paper 77-3 149pp $5.00

Lepawsky, Albert, ed.
The Prospect for Presidential-Congressional Government. 110pp $6.00

Nathan, Harriet and Stanley Scott, eds.
Emerging Issues in Public Policy: Research Reports and Essays 1973-1976. 164pp $11.00

Monographs, Bibliographies, Research Reports and a full list of Institute publications are available from the Institute of Governmental Studies, 109 Moses Hall, University of California, Berkeley, California 94720. Checks should be made payable to the Regents of the University of California. Prepay all orders under $30.00. California residents add 6% sales tax; residents of Alameda, Contra Costa and San Francisco counties add 6½% sales tax. Prices subject to change.